Revolutionary in Exile

The Emigration of Joseph Priestley to America

1794–1804

TRANSACTIONS

of the

American Philosophical Society

Held at Philadelphia for Promoting Useful Knowledge

VOLUME 85, Part 2, 1995

Revolutionary in Exile
The Emigration of Joseph Priestley to America
1794–1804

JENNY GRAHAM

THE AMERICAN PHILOSOPHICAL SOCIETY

Independence Square, Philadelphia

1995

Library of Congress Catalog
Card Number: 94-78516
International Standard Book Number 0-87169-852-8
US ISSN 0065-9746

For John
who has enjoyed Priestley, too.

"With my best wishes for the success of your laudable endeavours in the cause of *science, truth, justice, peace,* and, which comprehends them all, and every thing valuable in human life, LIBERTY, I subscribe myself

>Dear Sir,
>Your affectionate (?friend: torn)
>and most obliged
>humble Servant
>
>Joseph Priestley"

Priestley to (Franklin) 14 February 1769, A.L.S., American Philosophical Society, Franklin Papers, W. B. Willcox, ed., *The Papers of Benjamin Franklin*, XVI. (Yale University Press, 1972), 41–2.

"His antagonists think they have quenched his opinions by sending him to America, just as the pope imagined when he shut up Galileo in prison that he had compelled the world to stand still."

Jefferson on Priestley, J. Bernard, *Retrospections of America, 1797–1811* (New York, 1887, repr. 1969), 238.

"I will give you a specimen of his republicanism. 'The time,' said he to me, 'will I hope one day come when *laws* shall govern so completely that a man shall be a month in America without knowing who is President of the United States.'"

Benjamin Rush to Jefferson, 4 February 1797, L. H. Butterfield, ed., *Letters of Benjamin Rush* (*Memoirs of the American Philosophical Society*, Vol. 30, Part 2, Princeton University Press, 1951), II. 786.

-

TABLE OF CONTENTS

ILLUSTRATIONS

ABBREVIATIONS

Manuscript Sources

A. P. S.	American Philosophical Society
B. R. L.	Birmingham Reference Library
Bowdoin Coll.	Bowdoin College, Special Collections
Dickinson Coll.	Dickinson College, Special Collections
D. W. L.	Dr. Williams's Library
Lib. Cong.	Library of Congress
L. P. L.	Liverpool Public Libraries
Mass. Hist. Soc.	Massachusetts Historical Society
Penn. Hist. Soc.	Pennsylvania Historical Society
Royal Soc.	Royal Society
T. S.	Treasury Solicitor's Papers, Public Record Office
W. P. L.	Warrington Public Libraries

Periodicals

Amer. Hist. Rev.	American Historical Review
E. H. R.	English Historical Review
Mem. Am. Phil. Soc.	Memoirs of the American Philosophical Society
Proc. Am. Phil. Soc.	Proceedings of the American Philosophical Society
P. M. H. B.	Pennsylvania Magazine of History and Biography
P. M. H. S.	Proceedings of the Massachusetts Historical Society
Trans. Am. Phil. Soc.	Transactions of the American Philosophical Society
Trans. Royal Hist. Soc.	Transactions of the Royal Historical Society
W. M. Q.	William and Mary Quarterly

PREFACE

The author of any work attempting to describe the thought and achievements of Joseph Priestley must be indebted not only to the inspiration which its subject matter provides, but also to the help and cooperation of a very great number of people. Priestley's still voluminous extant correspondence, the quantity of his published output on any subject, and the recognition which was eventually to be his both in England and America, have ensured a large corpus of manuscript and printed material in libraries and other institutions on both sides of the Atlantic.

I am deeply grateful to all those who have assisted in the preparation of this study: to Elizabeth Carroll-Horrocks, of the American Philosophical Society, and to John Creasey, of Dr. Williams's Library, London, for their unfailing courtesy and helpfulness, and for permission to quote from the extensive Priestley material in their Collections. I am similarly indebted to the Earl of Shelburne, the Librarians of the Royal Society, Dickinson College, Warrington Public Libraries, Birmingham Public Libraries, the Historical Society of Pennsylvania, the Library of Congress, Liverpool Public Libraries, Massachusetts Historical Society, Bowdoin College Library, Haverford College Library, the New York Public Library, and the Public Record Office.

Many people also very kindly helped to provide the illustrations for this publication, and I am extremely grateful to Sandra Cumming, of the Royal Society, and Roy Goodman, of the American Philosophical Society, for their assistance and advice. I would also like to thank Diana Vaughan Gibson for her very great generosity in allowing the reproduction of the portraits of Priestley and Mary Priestley, and the portrait by Thomas Badger of Benjamin Vaughan; and also for her kind interest in my work.

This study has been read and commented upon by a number of scholarly advisers. I would like to express my appreciation to the anonymous readers of the American Philosophical Society; to Dr. Colin Bonwick, who most generously shared his knowledge of Priestley, and gave advice and encouragement; and to Dr. David J. Jeremy, for his helpful and constructive advice. Professor Merrill D. Peterson most kindly helped me to locate the source for the quotation from Jefferson on p. vi. To Professor Ian Christie, who has provided unstinting support and guidance throughout my research, I owe a debt of gratitude which can never adequately be repaid.

I should like to thank my Editor, Carole Le Faivre Rochester, for skillfully correcting the vagaries of my prose. As all who have had the privilege of working in the Library of the American Philosophical Society well know, it is a place of helpfulness and informed scholarship which can have few equals. The Library of Swarthmore College, Pennsylvania, has also been for me for many years a haven of retreat and repose. I would like to record my thanks for the kindness and hospitality that I have received there and also to the Libraries of Bryn Mawr, Haverford, and the University of Pennsylvania, for their courtesy and generosity in allowing me access to their Collections. I am more than grateful to the typesetters of this work, Asterisk in Barre, Vermont, for their patience and expertise.

My record of appreciation would not be complete without acknowledging the very considerable help and support which I have received from my family—in particular from John Graham, who has read and commented upon much of this work. Oliver Graham read the Introduction, and has been a source of much excellent advice; and both he and William Graham have given me stalwart, loyal and invaluable support. For them, as for me, I hope and believe, it has been a privilege to have spent time with Joseph Priestley. For the opinions expressed, and any errors incurred, I alone take responsibility.

JENNY GRAHAM
Lucy Cavendish College
Cambridge

Swarthmore, Pennsylvania

INTRODUCTION

In 1794, according to one contemporary estimate, some 10,000 persons, considerably more than in any preceding or subsequent years until the end of the Napoleonic wars, emigrated to the United States of America.[1] This influx of immigrants from the countries of Europe—in particular from England—has in the past been much neglected by historians, and indeed in one recent survey of the peopling of America accorded little more than a passing mention.[2] As one at least of the earlier historians rightly observed, however, "the 1790s beheld one of those migrations from Britain which, though so easily lost in the mass of Americans, has served to keep the United States in closer social and intellectual communion with the country from which it politically sprang."[3]

In several more recent studies, further substance has been given to the influence upon the political life of America of this "most respectable group" to arrive in the New World since the settlement of the New England colonies.[4] In the work of Arthur Sheps,[5] Richard Twomey,[6] and most recently Michael Durey, systematic studies have been undertaken emphasising the great number of emigrants involved; the active role many of them played in the English radical movement that developed in the French revolutionary era—which for many of them was the immediate cause of their flight to America; and the contribution which so many of these inveterate political propagandists made to the articulation of the issues at stake in the increasingly troubled politics of that country. "Marginal" in England, in America their contribution was based upon a recognition, in a more open society, of their talents. In underestimating the numbers of emigrants from England, in failing fully to appreciate the collective contribution which their standing as radical activists enabled them to make, historians, argues Durey, have largely overlooked a vital component in the emergence of the philosophy that came to be known as Jeffersonian Republicanism. In their articulation of a political and

---

[1] W. J. Bromwell, *History of Immigration to the United States* (New York, 1856, repr. Arno, 1969), 13–15.

[2] F. D. Scott, *The Peopling of America: Perspectives on Immigration* (Washington, American Historical Association, 1984), 20, 48.

[3] M. L. Hansen, *The Atlantic Migration, 1607–1860* (Harvard Univ. Press, 1940), 59. The monumental survey of immigration to North America by Bernard Bailyn, *Voyagers to the West* (New York, 1987), has to date surveyed the years prior to 1776.

[4] R. J. Twomey, "Jacobins and Jeffersonians: Anglo-American Radical Ideology, 1790–1810," in M. and J. Jacob, eds., *The Origins of Anglo-American Radicalism* (London, 1984), 284.

[5] A. Sheps, "Ideological Immigrants in Revolutionary America," in P. Fritz and D. Williams, eds., *City and Society in the Eighteenth Century* (Toronto, 1973), 231–46.

[6] R. J. Twomey, "Jacobins and Jeffersonians," above, n. 4.

economic philosophy in opposition to the Federalists, and in propagandising this philosophy as editors of newspapers in the Republican press, the English emigrés made a contribution which was influential in the shaping of opinion, and were accorded the recognition both of their allies and their opponents, commensurate with the part which they played.[7]

It is the purpose of the present study to examine, in considerably more detail than has previously been essayed, the career of one who was arguably the most prominent of all the political exiles from England at this time, the radical scientist, theologian, and political philosopher, Joseph Priestley. Priestley's own protestations of his lack of active involvement in the politics of America have been largely echoed by his biographers,[8] and effectively also by those studies attempting to account for the influence of emigrant English activists and the development of their ideas in America. His presence is given at best a passing mention, and his influence and ideas similarly subjected to no very rigorous analysis.[9] But in the circumstances of his departure from England, in his eventual heated involvement in the issues at stake in America under the administration of the Federalists, and in the recognition and approval accorded him by Jefferson, Priestley's years in America are worthy of study.

For several years before his enforced departure for America in April 1794, Priestley occupied a position in the intellectual and political life of England, and suffered a persecution from his opponents, which led many of his admirers to a comparison with Socrates. In the fields of science, religion, and politics, Priestley had aroused an admiration, and a corresponding antagonism, for his fearless questioning of established dogma, for his pursuit of the truth as he perceived it as a result of experimental investigation and enquiry, and for his willingness to propagate the principles to which he adhered in order to challenge existing orthodoxy. Priestley was a Unitarian Minister of religion who openly attacked

---

[7] M. Durey, "Thomas Paine's Apostles: Radical Emigrés and the Triumph of Jeffersonian Republicanism," *W. M. Q.*, 3rd Series, Vol. 44 (1987): 661–88; and also Durey, "Transatlantic Patriotism: Political Exiles and America in the Age of Revolutions," in C. Emsley and J. Walvin, eds., *Artisans, Peasants and Proletarians, 1760–1860* (London, 1985), 7–31.

[8] C. Robbins, "Honest Heretic: Joseph Priestley in America, 1794–1804," *Proc. Am. Phil. Soc.*, 106.1 (1962): 69: although cf. Professor Robbins's comment, ibid., 61, n. 12, that "Priestley exaggerated his indifference to politics and his lack of involvement in it both in England and America," and her commentary, ibid., 72–6, on his political writings in America. Cf. also C. Bonwick, "Joseph Priestley, Emigrant and Jeffersonian," *Enlightenment and Dissent*, 2 (1983): 3, although cf. Dr. Bonwick's further statement, ibid., 9, that Priestley was "much concerned with political behaviour as an integral component of his philosophy." His valuable analysis of the interest which, seen in this light, the experiment in government represented for Priestley, and the extent to which he commented upon its constitutional provisions and translated his social and economic ideas into the American context, has been strangely neglected by Durey (above, n. 7).

[9] Cf. Twomey, "Jacobins and Jeffersonians," above, n. 6; and also Durey, "Transatlantic Patriotism," 9, where he is concerned to correct what he sees as an overemphasis upon a few names, such as Priestley's. In his later study, "Thomas Paine's Apostles," 678–9, Durey tends, in his brief consideration of Priestley, to echo the tendency of many historians to place greater emphasis on the writings and influence of his close friend and political disciple, Thomas Cooper.

the Anglican establishment, and was a professed believer in the importance of instructing the young; a master of scientific experimentation, whose laboratory—in his own words, "the most truly valuable and useful apparatus of philosophical instruments, that perhaps any individual, in this or any other country, was ever possessed of"[10]—had enabled him to contribute to the burgeoning industrial and commercial development in particular of the midlands of England; and a political philosopher and polemicist who had enjoyed the closest of friendships with Franklin, and in the period of the French Revolution was a particular target of Burke.

In any consideration of Priestley, the intertwining of his many interests, and their interaction in the tendency of his thinking, must be borne in mind. The aim of the present study, however, is to concentrate as far as possible upon the political life of Priestley, and to carry forward into his ten years in America the themes set out by the author in a previous paper on his years in England.[11] The essentially radical and revolutionary nature of Priestley's political philosophy, the very great interest which, despite his many protestations to the contrary, it held for him, and the active part which on more than one occasion he played in the political arena, can, it is believed, be demonstrated further by a consideration of his years in America. Priestley arrived in America as one of the many "ardent Spirits in Europe," as John Adams described the refugees,[12] a symbol for many of the republican principles which he had done so much to propagate. In his articulation of the issues at stake in that country, and his adaptation of some at least of his political views to the needs of an emergent nation, his inveterate, invariably contentious, but also passionately committed interest in politics can very clearly be seen. It was in America, as he wrote after the election of Jefferson to the presidency in 1801, that he finally found himself "in any degree of favour with the governor of the country in which I have lived."[13] A study of his years there, in many ways as full of turmoil as were his years in England, can afford much further insight into the political personality of this remarkable and complex man. And insofar as he became drawn into and, it will be argued, to some extent influenced, the political scene in America, a consideration of his contribution can help to throw light upon the issues which dominated that decade, and which have been the subject of much recent historical debate. Before considering Priestley's career in America, however, some attention must be paid to the chief events of his years in England.

---

[10] Priestley, *Letter to the Inhabitants of Birmingham, Morning Chronicle*, 20 July 1791.

[11] J. Graham, "Revolutionary Philosopher: The Political Ideas of Joseph Priestley (1733–1804), Part One," *Enlightenment and Dissent*, 8 (1989): 43–68; and "Part Two," ibid., 9 (1990): 14–46.

[12] Adams to Jefferson, 11 May 1794, L. J. Cappon, ed., *The Adams-Jefferson Letters. The Complete Correspondence between Thomas Jefferson and Abigail and John Adams* (Chapel Hill, 1959), I.254–5.

[13] Priestley to Logan, 26 December 1801, Penn. Hist. Soc., Logan Papers, V.43.

Priestley was nurtured from an early age in the tradition of Protestant Dissent. His intellectual abilities were very evident; and his determination to become a nonconformist minister was interrupted only briefly by a period of ill health during which it was determined to apprentice him in the countinghouse of a merchant in Lisbon. In 1752, he entered the Dissenting Academy of Daventry, where he first encountered, and was profoundly influenced by, the associationist philosophy of Hartley. He was subsequently to minister to congregations in Suffolk and in Cheshire, and although, in 1761, in a step which was to have momentous consequences for his future development, he became a tutor at the Dissenting Academy of Warrington, it was also at this time that he was formally ordained. It was above all as a minister of religion that Priestley, in composing his *Memoirs*,[14] clearly wished to be remembered. The emphasis which he placed upon his activities in this sphere—an emphasis which was reiterated by his first biographer, Rutt[15]—has, as has been pointed out by R. E. Schofield in his authoritative account of Priestley's scientific work, had the effect of detracting from an appreciation of the full scope and originality of Priestley's achievements in the field of science.[16] So too the comparatively brief and at times even deliberately misleading account of his political activities by Priestley in his *Memoirs*, and by Rutt in his editing of his letters, has, as has been argued elsewhere, led to a severe underestimate of his deep commitment to and of his influence on the political movements of the age.[17]

It was at the Academy at Warrington that Priestley first came into close contact with many of the members of the dissenting elite of England, not only among his fellow tutors, but also in the families of those who were committed to his care. Prominent among these were the sons of Samuel Vaughan, the wealthy London West India merchant who was a committed opponent of the English ministry in the tumultuous politics embodied in the turbulent career of Wilkes, and an ardent partisan throughout this period in the cause of the American colonists. For several years after the achievement of American independence, Samuel Vaughan settled, with several members of his large family, in Philadelphia.[18] Four of Samuel Vaughan's sons were educated at Warrington, and the two elder, Benjamin and William, were, at his special request, placed under Priestley's immediate care—lodging in the house in which he now lived

---

[14] *Memoirs of Dr. Joseph Priestley . . . written by himself . . . by his son, Joseph Priestley: and Observations on his Writings by Thomas Cooper . . . and the Rev. William Christie* (Northumberland, 1806, repr. New York, 1978).

[15] J. T. Rutt, ed., *The Theological and Miscellaneous Works of Joseph Priestley* (London, 1817–1831), I. Parts 1 and 2: *Life and Correspondence.*

[16] R. E. Schofield, ed., *A Scientific Autobiography of Joseph Priestley (1733–1804)* (Cambridge, Mass., and London, 1966), vi–vii.

[17] Graham, "Revolutionary Philosopher, Part One," 48–9.

[18] Ibid., 52; and S. P. Stetson, "The Philadelphia Sojourn of Samuel Vaughan," *P. M. H. B.,* 73 (1949): 459–74.

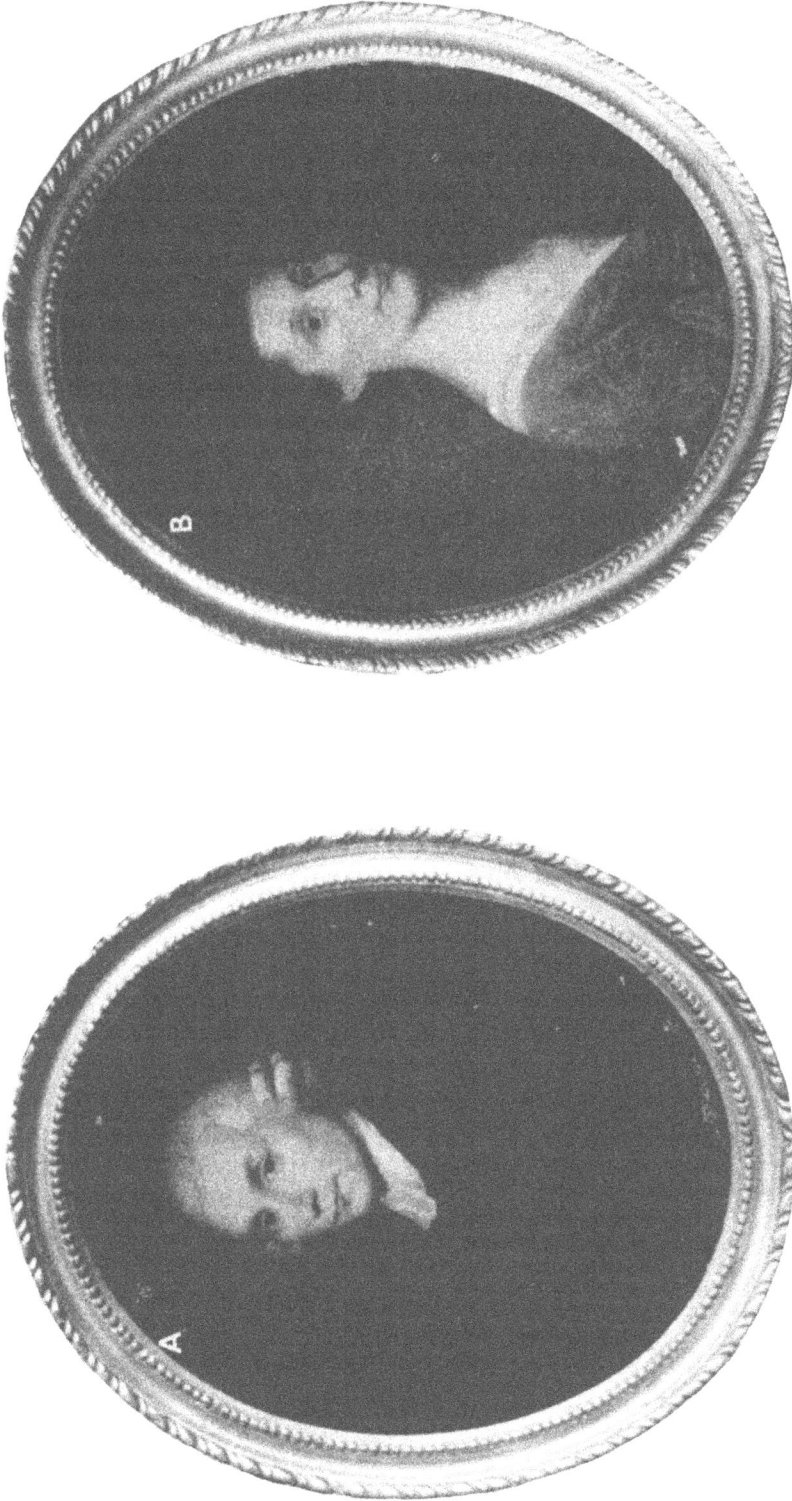

FIGURE 1. Joseph and Mary Priestley, 1760–1770, artist unknown. Courtesy of Diana Vaughan Gibson.

with his wife, the daughter of the ironmaster, Isaac Wilkinson.[19] Benjamin Vaughan's early fascination with politics can be seen in one little cited letter written while he was at Warrington, discussing the prospects of Wilkes. His admiration for Priestley is evident in a letter to Franklin: "You will find him great, though various; and as perfect a philosopher, as though he had never been a divine. . . . In his day, his works will have their effect, and some of them, for ever."[20] The regard in which Priestley held Benjamin Vaughan was testified to in his dedication to his former pupil of his *Lectures on History and General Policy*, based upon those which he had given at Warrington.[21] Their mutual intimacy with Franklin, and closeness of political interest, was made very evident with the inclusion by Benjamin Vaughan, in his edition of Franklin's writings – published in the midst of the strife of war, in 1779 – of three letters from Franklin to Priestley. Written on Franklin's return to America after his long sojourn in England, they testified in moving terms to his deep and bitter anger at the destructive effects of the policies of the English ministry, and his conviction that a separation from the mother country, and the sufferings of war, were now inevitable.[22]

In the winter of 1765-6, on a visit to London, Priestley had been first introduced to Franklin, and began a friendship which was to make him perhaps the closest of all Franklin's confidants in England. It was a friendship which was profoundly to influence Priestley's subsequent career, not least in the encouragement it afforded him in the field of science. For it was to Franklin that Priestley confided his wish to write a history of the current state of knowledge concerning electricity; and it was Franklin who, above all, encouraged Priestley in his endeavours. From this time onwards, Priestley's increasing interest in his own scientific experimentation, his assembling of as much scientific apparatus as he could procure, and his rapid publication of his results, earned him the recognition of the circle of scientists and philosophers in the capital, and his early election to the Royal Society.[23]

In the sphere of politics also, Priestley's contact with the circle of radical intellectuals in the metropolis was shortly to yield results. Early in 1768 he published his *Essay on Government*, and in the following year his *Present State of Liberty in Great Britain and her Colonies*, two pamphlets in

---

[19] *Monthly Repository*, IX. (1814): 267, 390, 530. T. E. Thorpe, *Joseph Priestley* (London, 1906), 50-1. For the influence which Priestley's relationship with the Wilkinsons was to have upon his career, cf. W. H. Chaloner, "Dr. Joseph Priestley, John Wilkinson and the French Revolution, 1789-1802," *Trans. Royal Hist. Soc.*, 5th Series, 8 (1958): 21-40.

[20] T. E. Thorpe, *Joseph Priestley*, 41; and Benjamin Vaughan to Franklin, n.d. (1767-8), Franklin Papers, A. P. S., cit. C. Murray, *Benjamin Vaughan (1751-1835). The Life of an Anglo-American Intellectual* (New York, 1982), 25 and note.

[21] Priestley, *Lectures on History and General Policy* (Birmingham, 1788, 2nd ed., Philadelphia, 1803), *Works*, XXIV.3.

[22] B. Vaughan, ed., *Political, Miscellaneous, and Philosophical Pieces . . . written by Benjamin Franklin* (London, 1779), 365-6, 550-54; Priestley, *Works*, XXV.392-3; and cf. B. Oberg et al., eds., *The Papers of Benjamin Franklin* (Yale Univ. Press, 1992, 1993), XXIX, XXX.

[23] Schofield, *Scientific Autobiography*, 12ff.

which, in enunciating his own political philosophy, and pronouncing upon the issues at stake between the English ministry and its opponents both at home and in America, Priestley first gave public expression to a political outlook essentially Lockeian in its insistence upon the right of an oppressed people to rebel against a tyrannical government, and extreme to the point of being revolutionary in its articulation of this cornerstone of the ideology of English radicalism. It was original in its formulation of the principle that the crucial issue in the government of men in society, and the yardstick by which all the actions of government must be judged, was the preservation of "the good and happiness of the members, that is the majority of the members of any state"; but it was eloquent also in its insistence on the recognition of the essential individuality and equality of men. "Every man," wrote Priestley in his *Essay* of 1768, "when he comes to be sensible of his natural rights, and to feel his own importance, will consider himself as equal to any other person whatever." The other striking characteristic of Priestley's political outlook—his optimistic and unshakable belief in the infinite capacity of men for self-improvement, and their attainment of a perfect state of happiness within society, which he asserted also most notably in the *Essay* of 1768—enabled him to reconcile with little difficulty any conflict there might appear to be between these two assumptions.[24]

In the Preface to his *Essay* of 1768, Priestley candidly admitted the "freedom" with which he had made "this defence of civil and religious liberty":

They who know the fervour of generous feelings will be sensible, that I have expressed myself with no more warmth than the importance of the subject necessarily prompted, in a breast not naturally the coldest; and that to have appeared more indifferent, I could not have been sincere.[25]

His "free sentiments" in the *Essay* were commented upon by his mentor, Franklin.[26] And in more than one letter written at this time, Priestley clearly recognised the extremes into which his outspoken publication of his views, and the violence of his attacks upon the ministry, were leading him.[27] *The Present State of Liberty in Great Britain and her Colonies* was indeed published anonymously. And in 1770, in a passage in a letter to

---

[24] Priestley, *An Essay on the First Principles of Government and on the Nature of Political, Civil, and Religious Liberty* (London, 1768); *The Present State of Liberty in Great Britain and her Colonies* (London, 1769), *Works*, XXII.1–144; 380–98; P. N. Miller, ed., *Joseph Priestley. Political Writings* (C. U. P., 1993), 1–127; 129–44. And see Graham, "Revolutionary Philosopher, Part One," 54–7; I. Kramnick, "Republican Revisionism Revisited," *Amer. Hist. Rev.*, 87 (1982): 646; D. O. Thomas, "Progress, Liberty and Utility: the Political Philosophy of Joseph Priestley," in R. G. W. Anderson and C. Lawrence, eds., *Science, Medicine and Dissent: Joseph Priestley (1733–1804)* (London, 1987), 79; Miller, loc cit., xviii–xxi.

[25] Priestley, *Essay, Works*, XXII.6–7; Miller, ed., *Political Writings*, 6.

[26] Franklin to Dubourg, 22 September 1769, W. B. Willcox, ed., *The Papers of Benjamin Franklin* (Yale Univ. Press, 1972), XVI.205.

[27] Priestley to Lucy Aikin, 13 June 1769, Royal Soc. Mss., Priestley MS. 654: Yates Memorial Volume; Priestley to Lindsey, 18 December 1769, Priestley, *Works*, I.1.105, cit. Graham, "Revolutionary Philosopher, Part One," 56–7.

Theophilus Lindsey[28] which Rutt in publishing it thought fit to omit, Priestley wrote of his recognition of the implications for his career in so freely speaking his mind. "I am fully convinced that, if I would make anything of my *philosophical work*, I must make the world believe what is by no means true, that I mind nothing else."[29] In 1774, however, at Franklin's behest, he wrote, but again anonymously, one further impassioned denunciation of the conduct of the English ministry, and defence of the cause of the Americans, in which his essentially republican political stance was made unmistakably clear.[30] In another letter to Lindsey, in 1771, his continuing and deep concern about the current state of political affairs, and his explicitly revolutionary outlook, is again very evident. "To me," he wrote, "everything looks like the approach of that dismal catastrophe described, I may say predicted, by Dr. Hartley, in the conclusion of his Essay, and I shall be looking for the downfall of Church and State together. I am really expecting some very calamitous, but finally glorious, events. Happy they who will be found watching in the way of their duty!"[31]

In two recent studies of Priestley's political philosophy, which have done much to restore a proper emphasis upon its importance and its at times quite extraordinary and revolutionary extremism, Priestley's interest in Hartley's interpretation of the prophecies, which was to continue throughout his career, has been accorded a crucial importance in the formulation of his political position, in the development of his ideas, and in the extremes to which he went in pressing them. "Priestley's nearly total acceptance of the ideology of the French Revolution had the effect of intensifying his millenarianism," wrote Clarke Garrett: "It is thus impossible to separate his political opinions from his religious convictions."[32] Jack Fruchtman, Jr., in his detailed analysis of the millenarianism of both Priestley and his close political and religious ally, Dr. Price, asserts that it gave to their republican thinking a tenor which distinguished it "fundamentally" from that of the other radicals of their day: "They relied heavily on prophetic wisdom in order to judge which political and individual transformations best prepared men for the end of

---

[28] Theophilus Lindsey (1723–1808), was a leading member of the Unitarian movement in England, and from this time onwards until Priestley's death, perhaps the most valued of all his many friends: *D. N. B.*; T. Belsham, *Memoirs of Theophilus Lindsey* (London, 1812); H. McLachlan, *The Letters of Theophilus Lindsey* (Manchester, 1920). And cf. Priestley's reference to Lindsey in one of the Sermons which he delivered shortly before leaving England in 1794: "one particular Christian friend, in whose absence I shall, for some time at least, find all the world a blank." Priestley, *Works*, XV.532, and below, n. 107.

[29] Priestley to Lindsey, 23 December 1770, ibid., I.1.127–8, and D. W. L., for the passage omitted by Rutt.

[30] Priestley, *An Address to Protestant Dissenters of all Denominations, on the Approaching Election of Members of Parliament* (London, 1774), *Works*, XXII.483–98; and cf. Graham, "Revolutionary Philosopher, Part One," 58–60.

[31] Priestley to Lindsey, 23 August 1771, *Works*, I.1.146.

[32] C. Garrett, "Joseph Priestley, the Millennium, and the French Revolution," *Journal of the History of Ideas*, 34 (1973): 57; and *Respectable Folly: Millenarians and the French Revolution in France and England* (Johns Hopkins Univ. Press, 1975), 133.

time"; and in considering the social problems of the day, "they were actually more interested in turning to the past to look at the old re-vealed prophecies to see how they were being fulfilled in their own day than they were in formulating an ideology of an emerging bourgeois society."[33]

Such an analysis, however, bears little relation to the true content and significance of Priestley's political ideas. Priestley's political philosophy was – as he himself attested – the result of close reading of the works of the great mentors of the English radical tradition, above all those of Locke.[34] It was grounded within this tradition of English libertarianism, and its essential premises and his far-reaching application of them rested, as Isaac Kramnick has rightly observed, on more than the interposition of divine providence.[35] Priestley's millenarian convictions were certainly given encouragement, as his letter of 1771 demonstrates, by the events leading up to the war with the colonies; they were subsequently, as Garrett's work makes clear, to be reinforced by the events of the early part of the Revolution in France. In this sense they can be said to have given added emotional force and perhaps some additional extremism to his political convictions; and they undoubtedly contributed, as Martin Fitz-patrick has suggested, to Priestley's frequently (and paradoxically) cavalier approach to political compromise and manoeuvre.[36]

Priestley's obsession with prophecy increased, moreover, as he him-self was the first to admit, in the isolation which was so frequently to be his lot in America. But that his millenarianism fundamentally affected the actual intellectual content of his political beliefs, and his application of them to the problems of society, or that it critically distinguished his thinking from that of other radicals, must be doubted. Moreover – as a corollary of this – that it in any way affected the influence of his political philosophy upon the radical thought of his age, must also be questioned. It was to Priestley, wrote Thomas Cooper – the English radical activist who was his close friend and political disciple in England and America, but who was himself a declared unbeliever – that an entire generation owed "the first plain, popular, brief and unanswerable book on the prin-ciples of civil government."[37] It was not Priestley's millenarianism as such which his enemies were to fasten upon, and his many disciples to admire. It was his intellectual grasp of the problems of government, his essentially rational, if frequently extreme, approach to their solution, and

[33] J. Fruchtman, Jr., "The Apocalyptic Politics of Richard Price and Joseph Priestley: A Study in Late Eighteenth-Century English Republican Millennialism," *Trans. Am. Phil. Soc.*, 73, 4 (1983): 34, 44, 46, 55, although cf. ibid., 63.

[34] Priestley to Thomas Hollis, 1 November 1768, *Works*, I.1.94–5, cit. Graham, "Revolu-tionary Philosopher, Part One," 53.

[35] Kramnick, "Republican Revisionism Revisited," 646; "Eighteenth Century Science and Radical Social Theory: The Case of Joseph Priestley's Scientific Liberalism," *Journal of British Studies*, 25.1 (1986): 14.

[36] M. Fitzpatrick, "Joseph Priestley and the Millennium," Anderson and Lawrence, eds., *Science, Medicine and Dissent*, 34–5.

[37] Priestley, *Memoirs*, II.354–5.

his ability, as Thomas Cooper also wrote, "as a writer on the theory of politics, a subject in which the development of a simple truth in such a manner as to impress it on the mind of the public may influence, to a boundless extent, the happiness of millions."[38]

From 1774, and throughout the years of the war with America, Priestley's political sympathies were never in doubt. His writings, with those of Price, were of considerable influence in the formulation of the opposition to the English ministry in America; and he was remembered in that country not only by Franklin but by others, as one who remained "unterrified and unseduc'd from the cause of truth and Liberty."[39] It was at this time however, that Priestley most strikingly avoided political polemics and public political controversy. In 1773 he had accepted the position of librarian to Shelburne and, as he later wrote, "entered into almost all his views, thinking them just and liberal." It was at Bowood that Priestley made many of his most important scientific discoveries, and he was also undoubtedly privy to much political discussion and debate. His relationship with his patron was, nevertheless, to be soured and eventually dissolved as a result of what Benjamin Vaughan in a letter to Franklin perceptively described as a lack of mutual understanding: "The one did not comprehend enough the nature and merit of a speculative scholar, nor the other the situation and difficulties of a political actor. I labored, as you did, to prevent it."[40] In 1779 Priestley wrote to Franklin of the possibility of reviving "a scheme" which Franklin had once suggested for him—that of an academic post at a College in America.[41] In 1780, however, in what he himself was to describe as "the happiest event in my life, being highly favourable to every object I had in view, philosophical or theological," Priestley moved to Birmingham, where he was immediately invited to become one of the ministers at the Unitarian New Meeting.

It was in Birmingham that Priestley immersed himself in theological controversy, and became a leading member of the most celebrated of all the English provincial intellectual societies, the Lunar Society, whose members were united in a common credo of faith in the advancement of science, and, as Isaac Kramnick has so brilliantly demonstrated, in its role in the advancement of the cause of social and political reform.[42]

---

[38] Ibid., II.337; and for Jefferson's rejection of Priestley's millenarianism, see A. Koch, *The Philosophy of Thomas Jefferson* (New York, 1943), 34–5.

[39] Arthur Lee to Price, 20 April 1777, *P. M. H. S.* (1903): 310, cit. Graham, "Revolutionary Philosopher, Part I," 61; C. Bonwick, *English Radicals and the American Revolution* (Chapel Hill, 1977), 78, 107–8, 152; Kramnick, "Republican Revisionism Revisited," 639ff.

[40] Priestley, *Works*, XV.525, and below, n. 107; B. Vaughan to Franklin, 26 June 1780, Franklin Papers, A. P. S., cit. Murray, *Benjamin Vaughan*, 60; and cf. "Revolutionary Philosopher, Part One," 60.

[41] Priestley to Franklin, 11 March 1779, B. Oberg et al., eds., *Papers of Benjamin Franklin*, XXIX.99, and note.

[42] Priestley, *Memoirs*, I.97; R. E. Schofield, *The Lunar Society of Birmingham; a Social History of Provincial Science and Industry in Eighteenth Century England* (Oxford, Clarendon, 1963); Kramnick, "Joseph Priestley's Scientific Liberalism," 6ff.

FIGURE 2. Joseph Priestley, engraving by William Bromley, published in the *European Magazine*, 1 January 1791. By permission of the President and Council of the Royal Society.

Science, Priestley had written, citing Bacon, in his *Essay* of 1768, could make the "situation of men in this world abundantly more easy and comfortable."[43] The Lunar Society included within its ranks and was in correspondence with some of England's most prominent manufacturers and industrialists: Josiah Wedgwood, Matthew Boulton, James Watt, William Withering, James Keir, and many others. With the aid of a generous subscription from many of those who benefitted from his researches, Priestley laboured in his laboratory "with," as he later wrote, "no pecuniary view whatever, but only in the advancement of science, for the benefit of my country and of mankind."[44]

In 1788, in the midst of these years when he was closely associated with many of England's leading manufacturers and industrialists, Priestley published his *Lectures on History and General Policy*. And in his chapters on the place of economic activity within society and its relationship to government, he endeavoured to assimilate the views and interests of commerce and the manufacturers with whom he very clearly identified, with his strongly held convictions of the role of government in society. Agriculture, he wrote, in a passage which owed much to the author of the *Wealth of Nations*, was the fundamental key—"the only stable foundation"—to human activity and prosperity, without which commerce could not flourish.[45] It was however itself essentially dependent upon the stimulus provided by commerce: "The advantages of *agriculture* and *commerce* are reciprocal." And as a field of human activity, commerce, wrote Priestley, did much, not only to enrich the nation, but to enlarge men's minds, to inculcate the virtue of punctuality, and with it "the principles of strict justice and honour." If motives of "sordid avarice" might affect small concerns, they seldom affected those of larger—rather, the contrary: "By commerce numbers acquire both the wealth and the spirit of princes." Commerce flourished best, and the spirit of industry and the need for peace between nations which it encouraged, was best promoted, when it was least interfered with by government. "Of all the classes of men above mentioned," wrote Priestley, citing Adam Smith, "the governors are, in general, the most ignorant of their own business, because it is exceedingly complex, and requires more knowledge and ability than they are possessed of; though this is in consequence of their undertaking more than is necessary for the good of the state. If more was left to the attention and efforts of individuals, the business of government would not be so complex, and persons of inferior abilities might be equal to it."[46]

In these Chapters Priestley might indeed, as Isaac Kramnick has sug-

---

[43] Priestley, *Essay on Government, Works*, XXII.9.

[44] Schofield, *The Lunar Society*, 193ff; Priestley, *Letter to the Inhabitants of Birmingham, Morning Chronicle*, 20 July 1791, Priestley, *Works*, XIX.540-2.

[45] A. Smith, *An Inquiry into the Nature and Causes of the Wealth of Nations*, R. H. Campbell and A. S. Skinner, eds. (Oxford, Clarendon, 1976), I.426-7.

[46] Priestley, *Lectures on History and General Policy, Works*, XXIV.300-26.

gested, be included among those whom Burke derided for viewing the state as "nothing better than a partnership agreement in a trade of pepper and coffee, calico or tobacco, or some other such low concern."[47] His view of government, however, as having, with the laws, "the greatest influence on human affairs," was in implicit contradiction to the more extreme interpretation of such a view of society. "Of all the things which contribute to the domestic happiness and security of states, GOVERNMENT, with the various forms of it" was, he wrote elsewhere in the *Lectures*,

the first that offers itself to our notice. . . . The principle which leads men to form themselves into those larger societies which we call *states*, is the desire of securing the undisturbed enjoyment of their possessions. Without this the weak would always be at the mercy of the strong, and the ignorant of the crafty. But by means of government the strength and wisdom of the whole community may be applied to redress private wrongs.

And in his discussion of the resolution of the tensions which must always exist between individual aspirations and the needs of the community, and of the proper role of government in regulating the affairs of its citizens, he remarked that "governments, and systems of laws adapted to them, are more *simple or complex* according to the variety and connexion of the interest of the members of the community." In hunting and pastoral communities, and in societies based upon agriculture, little or no control was needed in order to regulate the affairs of men. In societies "addicted to commerce," however,

the connexions of individuals are the most intimate and extensive, and consequently their interests the most involved, that any situation of human affairs can make them . . . the smallest part of so complex a machine, as their government must necessarily be, has a variety of connexions, and the most important effects, and therefore requires to be adjusted with the utmost care.[48]

The principles of government set out in his *Lectures* were, Priestley declared, the same as those now taught in the newly founded Dissenting Academy at Hackney, and, "as I am informed, in the colleges in North America."[49] He paid at this time little attention to, and indeed, as he assured Benjamin Vaughan, did "not pretend to have any *opinion*" on the every day occurrences of politics; although, he added, "I like to hear what is passing, and what is thought of public measures by those of my friends whose opinions have weight with me."[50] It was in 1786, however, on one of his annual visits to London, that he first became acquainted with one of the most influential of the Americans in Europe, John Adams, while the latter was serving his term as ambassador to

---

[47] Kramnick, "Joseph Priestley's Scientific Liberalism," 20.

[48] Priestley, *Lectures*, *Works*, XXIV.220–9; cf. Kramnick, 18–21.

[49] Priestley, *Familiar Letters, Addressed to the Inhabitants of Birmingham* (Birmingham, 1790), *Works*, XIX.214.

[50] Priestley to Benjamin Vaughan, 2 March 1787, H. C. Bolton, ed., *Scientific Correspondence of Joseph Priestley* (Privately printed, New York 1892, repr. 1969), 82–3.

FIGURE 3. The destruction of Priestley's house and laboratory on 14 July 1791. Engraved from an oil painting, from a drawing made on the spot by Eckstein. By permission of the President and Council of the Royal Society.

FIGURE 4. Priestley's house and laboratory at Fairhill after the riots.
Engraving, from *View of the Ruins of the Principal Houses destroyed during the Riots at Birmingham*, 1791. By permission of the President and Council of the Royal Society.

England.[51] His own influence in England was at this time at its height, and was well described by James Currie in 1788:

Priestley, who spent an evening with me lately, is of opinion that no effort even of the humblest individual is ever lost. Let there be but agitation of any question, and the interests of truth and virtue are promoted; – no matter in what direction the motion comes – let there *be* motion, that is enough: the tide-mill goes equally whether the water runs with the flood or the ebb. This is a great man, and a most agreeable one.[52]

Above all in his years in Birmingham, Priestley dedicated himself to his theological concerns: the publication of his *History of the Corruptions of Christianity*, and an impassioned defence of the cause of religious toleration which, although fully in accord with developments across the Atlantic,[53] in England roused the full ire of the church establishment against him. In 1785 Priestley published his Sermon on *The Importance and Extent of Free Enquiry in Matters of Religion*, and, in a phrase which made full use of metaphorical scientific imagery, declared that the Dissenters were "laying gunpowder, grain by grain, under the old building of error and superstition."[54] And when in 1787, the Dissenters organised themselves to demand from the English government a repeal of the Acts which denied them their full rights as citizens, in forbidding them access to the corporations of the towns in which they were so numerous, and appointments to offices of state, Priestley, in a public *Letter to Pitt*, justified their aims, and declared once more that he and many others of his persuasion were avowed enemies of a church establishment.[55] In the campaign of 1789–90, in which the Dissenters, moved now by the powerful example of France, once more demanded of the English legislature a redress of their grievances, Priestley in his *Familiar Letters* once again – and to the dismay of many even among his co-religionists – argued the case for disestablishment.[56]

With the outbreak of Revolution in France, Priestley's political activities in England entered a new phase. He was greatly moved by the *Discourse* which Price delivered at the Annual Meeting of the London Rev-

---

[51] C. F. Adams, ed., *The Life and Works of John Adams* (Boston, 1856), III.396, 397; L. J. Cappon, ed., *The Adams–Jefferson Letters*, II.333–4; C. Bonwick, *English Radicals and the American Revolution*, 171ff; Z. Haraszti, *John Adams and the Prophets of Progress* (Harvard Univ. Press, 1952), 280; and Priestley, *Works*, XVI.3.

[52] W. W. Currie, *Memoir of the Life, Writings and Correspondence of James Currie* (London, 1831), I.136.

[53] J. G. A. Pocock, "Religious Freedom and the Desacrilization of Politics: From the English Civil Wars to the Virginia Statute," in M. D. Peterson and R. C. Vaughan, eds., *The Virginia Statute for Religious Freedom. Its Evolution and Consequences in American History* (C. U. P., 1988), 47.

[54] Priestley, *The Importance and Extent of Free Enquiry in Matters of Religion. A Sermon preached before the Congregations of the Old and New Meetings of Protestant Dissenters at Birmingham, Nov. 5, 1785. To which are added, Reflections on the Present State of Free Enquiry in this Country* (Birmingham, 1785), 40–1; and *Works*, XV.70–82.

[55] Priestley, *A Letter to the Rt. Hon. William Pitt . . . on the Subjects of Toleration and Church Establishments* (London, 1787), *Works*, XIX.111–34.

[56] Graham, "Revolutionary Philosopher, Part II," 16.

olution Society, and he was among the first to publish a reply to Burke's attack upon it in the *Reflections*. He welcomed with enthusiasm the publication of the First Part of Paine's *Rights of Man* in February 1791; and in a *Discourse* which he himself shortly afterwards delivered at the Dissenting Academy at Hackney, and in his active proselytising on behalf of a newly forming Constitutional Society in Birmingham, he revealed himself prepared to declare more publicly than for many years the political philosophy to which he had long adhered. "The generality of governments have hitherto been little else than a combination of *the few* against *the many*," he wrote in the conclusion of his *Letters to Burke*,

and to the mean passions and low cunning of these few, have the great interests of mankind been too long sacrificed. . . . How glorious, then, is the prospect, the reverse of all the past, which is now opening upon us. . . . Government, we may now expect to see, not only in theory and in books, but in actual practice, calculated for the general good, and taking no more upon it than the general good requires. . . . After the noble example of *America*, we may expect, in due time, to see the governing powers of all nations confining their attention to the *civil* concerns of them, and consulting their welfare in the present state only; in consequence of which they may all be flourishing and happy.

"In this new condition of the world," Priestley added, "there may still be *kings*, but they will be no longer *sovereigns*, or *supreme lords*, no human beings to whom will be ascribed such titles as those of *most sacred*, or *most excellent majesty*."[57]

In the spring of 1791, Priestley, after consultations with his manufacturing friends in Manchester—Thomas Cooper among them—had begun to establish the Constitutional Society which, with many others throughout England, intended to celebrate on 14 July the fall of the Bastille. Not only in Birmingham but in Manchester also was there much indignation at this public demonstration of sympathy with the radical and revolutionary new order of government in France. It was in Birmingham, however, whose leading Dissenters and manufacturers had become so closely associated with the very extreme religious and political views of Priestley, that the most terrible destruction occurred, and Priestley himself, as many believed, narrowly escaped with his life. In the "three days of uninterrupted mob violence" from 14–17 July 1791 in Birmingham, two Unitarian Meeting Houses were destroyed, the properties of some fourteen of the town's most prominent merchants and manufacturers were razed to the ground, and many others were threatened.[58] And in this appalling debacle it was Priestley, who in many ways personified the ideals and ambitions of this prosperous merchant class, who lost—to the ill-concealed satisfaction of his many enemies—his manuscripts, his books, his correspondence, and all of his philosophical apparatus.

---

[57] Priestley, *Letters to the Rt. Honourable Edmund Burke, Occasioned by his Reflections on the Revolution in France* (Birmingham, 1791), *Works*, XXII.237–8, 241.

[58] R. B. Rose, "The Priestley Riots of 1791," *Past and Present*, 8 (1960): 68–88.

J. PRIESTLEY, LL.D. F.R.S.

FIGURE 5.   Joseph Priestley, engraving by Angus, published in the *Literary Magazine*, 1 February 1792. By permission of the President and Council of the Royal Society.

FIGURE 6. Thomas Cooper (1759–1839) by Charles Willson Peale. Courtesy of the College of Physicians of Philadelphia.

PRIESTLEY'S DECISION TO EMIGRATE
TO AMERICA
JULY 1791–APRIL 1794

In the immediate aftermath of the rioting in Birmingham, Priestley, conscious of his immense unpopularity, and of the strong likelihood of further attacks upon his person, very soon considered leaving England. "It is now evident, from a variety of circumstances," he wrote to his brother-in-law Wilkinson,

that government is not displeased with the riots in Birm. Some of my friends who had occasion to wait on Mr. Dundas, say he did nothing but rail at the disaster in general, and myself in particular. . . .

On this account I consider my stay in this country as very uncertain. Many of (my) friends seriously think of going to France, and the neighbourhood of Dijon in Burgundy has been pointed out to them as convenient for their manufactures. If this should take place, and my son William get a settlement in France, which I hope my friends there, will find for him, I shall probably go too. Joseph says that many dissenters will probably emigrate from Manchester and that if all be well, he will be able to go too in a few years to great advantage.[59]

From France in the aftermath of the riots, which had been reported in the *Moniteur* on 27 July, had come addresses of condolence to Priestley, from learned societies, from the Jacobins in Paris and from Jacobin clubs in more than one provincial city, and also "very handsome proposals" of accommodation if he would remove to France—of a "house completely furnished" near Paris, and the prospect too of a vacant monastery near

---

[59] Priestley to Wilkinson, n.d. (September 1791), W. P. L.; and cf. W. H. Chaloner, "Dr. Joseph Priestley, John Wilkinson and the French Revolution," 27, where, however, there is an omission in the final sentence as quoted.

William Priestley had been in Paris in the summer of 1789; at the time of the riots, he was in his parents' household—about to be "three years with Mr. Russell, in order to his being afterwards settled in America." "He exposed himself much in the riots, in saving what he could of our things," Priestley wrote, "and was so marked by the rioters as to be in much danger." With Priestley's decision not to return to Birmingham, it was decided that it would be "an uncomfortable place" also for William. In June 1792 Priestley wrote that he was "to be in a merchants counting house, at Nantes, tho with a view to a partnership, if they should on trial agree." Priestley to Wilkinson, 20 August 1791, W. P. L.; Priestley to J. Vaughan, 7 June 1792, A. P. S., Priestley Papers, B. P. 931; and cf. also Priestley, *An Appeal to the Public on the subject of the Riots in Birmingham*, Part II (London, 1792), *Works*, XIX.506. For William Priestley's French citizenship in the summer of 1792, cf. below, n. 72.

Joseph Priestley, Jr. had in the spring of 1791 been settled with a merchant in Manchester: "Revolutionary Philosopher, Part II," 21, n. 23. But Ashworth, although recommended by all Priestley's friends in Manchester as a man of "liberal principles" was in the winter of 1792-3 to insist that Joseph Priestley, Jr., leave his firm immediately (cf. below, n. 90). For the part which young Priestley played in the radical politics of Manchester, his friendship with Thomas Walker, Thomas Cooper, and James Watt, Jr., and his own strong political convictions, cf. below, nn. 90, 190; and also Chaloner, 26, n. 4.

Toulouse, "which reason has recovered from superstition."[60] Priestley's sympathy with the politics of France was at this time at its height. And it was now that he began, apparently under the influence of Benjamin Vaughan (who was in France on several occasions in 1789–91 observing the progress of the Revolution)[61] to invest in the French funds, hoping, as he wrote, to reap as much advantage from them as he had done from his previous investments in America.[62] He continued, however, as is clear from his letters to another of the Vaughan family—John Vaughan, who had settled as a merchant in Philadelphia[63]—to invest there also.[64] "I wish to have as little in this country as possible," he wrote. "I am told

---

[60] Priestley, *Works*, I.2.127–9; 130–2; M. L. Kennedy, *The Jacobin Clubs in the French Revolution. The First Years* (Princeton, 1982), 240–1; Priestley to Thomas Wedgwood, 18 October 1791, *Scientific Correspondence of Priestley*, 116–17; Priestley to Wilkinson, 4 October 1791, W. P. L.; Priestley to J. Vaughan, 15 November 1791, A. P. S., Priestley Papers, B. P. 931; "Revolutionary Philosopher, Part II," 35–42.

[61] For Benjamin Vaughan's enthusiastic endorsement of the revolution in France, and for his reports to his patron, Lansdowne, on the political state of that country, cf. the author's forthcoming *Reform Politics in England, 1789–99*; and Murray, *Benjamin Vaughan*, 248ff.

[62] Chaloner, 28: citing Priestley's letter to Wilkinson, 4 October 1791, omitting, however, some words in the second sentence. This should read "He [Vaughan] has already placed a considerable sum in the French funds, and many, I doubt not, will soon do the same, as was the case with the American funds, which have risen thirty per Cent since I placed what I could in them. Mr. Russell got 30 per Cent per An by some money that he happened to have in their funds at a very critical time." In the event, Benjamin Vaughan placed £500 in the French funds at Priestley's request; and Wilkinson purchased £5,000 for him in the autumn of 1791, and a further £5,000 at some time between that date and the autumn of 1793 (Chaloner, 27–8).

[63] John Vaughan (1756–1841) was the fourth of the six sons of Samuel Vaughan, younger brother of Benjamin and William, and like them, educated at Warrington Academy. In 1782 he had emigrated to America and settled permanently in Philadelphia. As a longstanding friend of the family, Priestley had been among those who wrote testimonials for John Vaughan in 1782, and that Vaughan certainly knew Priestley well is suggested by his reaction to the news of the Birmingham riots (below, n. 67). John Vaughan had been elected a member of the American Philosophical Society in 1784, and it was perhaps partly due to his influence and of Samuel Vaughan, then Vice-President, that in 1785 Priestley was elected a member. By 1791 John Vaughan was Treasurer of the Society. In September 1791 he presented "a profile in Plaster of Paris of Dr. Priestley particularly valuable for the resemblance." From the summer of that year onwards he was in constant communication with Priestley, and closely involved in his plan of emigration. (S. P. Stetson, "The Philadelphia Sojourn of Samuel Vaughan," E. M. Geffen, *Philadelphia Unitarianism, 1796–1861* (Univ. of Pennsylvania Press, 1961), 20, 22–3; G. Chinard, "The American Philosophical Society and the World of Science, 1768–1800," *Proc. Am. Phil. Soc.*, 87 [1944]: 4; *Early Proceedings of the American Philosophical Society . . . compiled . . . from the Manuscript Minutes of its Meetings from 1744 to 1838* [Philadelphia, 1884], 196: 16 September 1791; Schofield, *Scientific Autobiography*, 243–4. And cf. also Franklin to R. Morris, A. P. S., Vaughan Papers, B. V. 462, containing a testimonial from Priestley to the character of John Vaughan, 5 January 1782.)

[64] Priestley to J. Vaughan, 22 October, 15 November, 7 December 1791, A. P. S., Priestley Papers, B. P. 931: although the yield on the American funds, which had now risen in value, was less, yet, he wrote, after a series of contradictory letters giving Vaughan instructions, he thought that "the *security* to be better than that of any European funds whatever." It seems clear, moreover, that in the summer of 1791 Priestley was adding to his investments in America: cf. J. Vaughan to Priestley, 2 August 1791, A. P. S., Vaughan Papers, B. V. 462.1: "Our heads in office who are acquainted with you[r] confidence in their administration & in the resources of the Country are much pleased with the circumstance. The approbation of those whose opinions we respect is truly valuable & makes us regardless of the Sneers of the ignorant and malicious."

it is the wish of the ministry to drive me away, and in this we shall soon be agreed." Within a month of writing in this vein, however, he was writing of his "absolutely" taking a house at Hackney; of his plans to refurbish his library and laboratory, and of his hopes of succeeding Price at the Gravel Pit Meeting in Hackney, although, as he wrote, "some of the more timid part of the congregation" were "apprehensive of a tumult" should he "settle there." In France, he had decided, he would be "useless. I shall therefore abide the storm, whatever it be. I cannot suffer much more than I have done." And yet, as he wrote in this same letter to John Vaughan: "I do not think it probable that I shall continue here many years, if my life be preserved." He intended, he wrote, to settle his sons in France or America, and at present "a favourable situation seems more likely to offer in France; and wherever they settle, I shall, as I think at present, finally go to end my days."[65]

To Wilkinson it seemed astonishing that Priestley did not take advantage of such "flattering invitations to remove into a Situation of Safety compared with that you are in on this Side the Water. Was I in your case," he wrote, "I should not hesitate long where to fix." And, he added, in regard to the spirit of bigotry abroad in England, which seemed determined to have Priestley "destroyed by any Means," and which would "not affect you so readily in its operations in France as it may in this Country . . . I should join in opinion with the *timid* part of your intended Congregation." And that Priestley himself, throughout 1792, concerned but increasingly discouraged in his attempt to gain some compensation for the losses he and his friends had suffered, and apprehensive of further rioting, regarded France as his natural refuge, is clear from his correspondence.[66] By the spring of 1792, however, he must certainly have received the letter from John Vaughan from Philadelphia, written shortly after the news of the Birmingham riots reached America, assuring him of his sympathy, and of the hopes of many in America that if the English government persisted in "their ill-judged encouragement of illiberal and unmanly sentiments . . . they will have to lament the loss & we to falicilitate (sic) ourselves upon the accession of a considerable number of the most enlightened liberal and industrious of her Citizens."[67] And in

---

[65] Priestley to Wilkinson, n.d. (September 1791), 4 October 1791, W. P. L.; Priestley to J. Vaughan, 15 November 1791, A. P. S., B. P. 931.

[66] Wilkinson to Priestley, 10 October 1791, W. P. L., Priestley to Lavoisier, 2 June 1792; Priestley to Withering, 2 October 1792; Priestley to Mrs. Crouch, 31 December 1792, *Scientific Correspondence*, 129–32 and note. Cf. Priestley to J. Vaughan, 7 June 1792, A. P. S., B. P. 931: "The populace, in many parts of England, are ready for the greatest outrages, on any pretext whatever. There has been a second riot at Birmingham, tho not of the same kind with the former, but it was quelled by the soldiers. More disturbance is expected there. There are also symptoms of the same spirit in London, and both parties look forward with more or less dread to the 14th of July." And cf. Priestley, *Memoirs*, I.119; *Works*, I.2.170–1, 176, for his not undisputed election at Hackney.

[67] J. Vaughan to Priestley, 3 October 1791, A. P. S., Vaughan Papers, B. V. 462.1 and Appendix; and cf. also J. Vaughan to B. Vaughan, 3 October 1791, ibid.: "I shall not be Surprised if emigrations are consequent upon the Countenance given in England to illiberal & intollerant Sentiments religious and political. I can only say we shall be ready to receive the chosen band with open arms."

Priestley's correspondence with John Vaughan at this time, discussing his investments in the American funds, and also Vaughan's offer to take one of his sons under his care, there is an increasing emphasis on the possibility of one at least of his sons settling in America. For Priestley himself, however, emigration remained "a distant tho a pleasing speculation."[68] "I approve your resolution of retiring from the scene for a time," he wrote to his close friend William Russell[69] in the summer of 1792, "though the idea of your final emigration is more than I can well bear, so intimately and happily connected as we have been. I wish it would suit me to accompany you, but to that there are (those) who would never consent."[70] He was in many ways, as he wrote, comfortable in his new situation at Hackney: "It is a most agreeable circumstance attending it," he wrote to Vaughan, "that so many of your family will be of the congregation."[71] And he was still, perhaps to a greater degree than ever, interested in the politics of revolutionary France.

In the summer of 1792 Priestley, on hearing of the naturalisation of his son William "from the public papers," as he wrote to Russell, nevertheless defiantly defended his son's actions, and those of the French:

I had no expectation of any such thing; but if it had been my own wish and procurement, what harm was there in it? This country is not likely to be a desirable situation for any child of mine, and therefore it is natural for me to look for a settlement for them elsewhere. On the other hand, it is natural for the people of Birmingham to be offended at whatever throws a reflection upon them, and they must expect much more exasperation of the same kind.[72]

---

[68] Priestley to J. Vaughan, 27 February, 7 June 1792, A. P. S., Priestley Papers, B. P. 931: "Being at this distance, I can only give you general directions, and having perfect confidence in your judgement, and willingness to serve me, I desire that you would, in whatever manner you think best, make the most of the two thousand pounds that have been remitted to you. . . . It is very probable that some time hence one of my sons may settle in America, and then I shall transfer the whole to him. . . . I shall . . . be as well satisfied if you think it to be to my advantage to make a purchase in *land*, in preference to buying any more stock." And, on Vaughan's offer to look after one of his sons: "You very obligingly mention your readiness to take a son of mine under your care. I wish you would be more specific on the subject."

[69] William Russell (1740–1818), was a prosperous Birmingham manufacturer, a member of the New Meeting, and active in the cause of political and religious reform in England. He was one of the chief organisers of the Bastille Day Dinner and his house was among those destroyed in the riots. He was particularly active in attempting to procure some redress from the ministry, but retired from Birmingham to Gloucestershire before his emigration to America in the summer of 1794 (*D. N. B.*; S. H. Jeyes, *The Russells of Birmingham in the French Revolution and America, 1791–1814* [London, 1911]; and also "Revolutionary Philosopher, Part II," 28–9).

[70] Priestley to Russell, 12 June 1792, *Works*, I.2.183; and cf. also Priestley to Vaughan, 22 October 1791, in reply to a letter from Vaughan "on the subject of a Unitarian minister" (cf. J. Vaughan to B. Vaughan, 3 October 1791, B. V. 462.1): "I . . . much wish that a proper preacher could be found," Priestley wrote: "But America itself must find the man, as in due time I doubt not will be the case": A. P. S., Priestley Papers, B. P. 931.

[71] Priestley to J. Vaughan, 7 December 1791, A. P. S., B. P. 931; and cf. Priestley to Mrs. Crouch, 31 December 1792, *Scientific Correspondence*, 132.

[72] Priestley to Russell, 22 June 1792, *Works*, I.2.185–6; and "Revolutionary Philosopher, Part II," 37.

Such thinking led him to accept, without apparent hesitation, the offer of citizenship for himself from the National Assembly in August, after the overthrow of the monarchy. And his sympathy with France, albeit mingled as it clearly was with apprehensions for its viability as a safe refuge, remained constant throughout these years. He was in close contact at this time with the English radical merchant, John Hurford Stone, who had been instrumental in securing for him the ministry at the Gravel Pit Meeting in Hackney, and who was, until his departure for France in April 1792, a neighbour of whom Priestley saw much. From Paris in the spring and late summer of 1792 Hurford Stone transmitted to his circle of friends in England a regular series of letters, giving close and detailed information on the fluctuating fortunes of the Revolution—for which his own strong sympathy was never in doubt. In August 1792, writing to his brother of the inevitable downfall of the monarchy, Hurford Stone wrote also to Priestley, and it was in response to that letter that Priestley entrusted Stone with a copy of his letter to the National Assembly, accepting their offer of citizenship, and revealing the close interest which he was also taking in the politics of France.[73]

In October Priestley was writing to Russell of his hopes that "the aspect of things in France will be clearing up," for, as he said, "much depends upon that, in every case in which civil or religious liberty is concerned." He gave Russell the latest news from France, brought to him at that very moment by "Mr. Vaughan," adding in a postscript that he had just seen "a letter from Mr. Stone, dated 1 o'clock, 2nd Oct., Hall of the Convention. This moment the news is arrived that the Prussians have raised their camp and are in flight."[74] Throughout the autumn of 1792, Priestley shared the widespread elation in English reforming circles as the armies of the sans culottes marched into Flanders. "The success of the French," he wrote to Rev. Edwards, of Birmingham, ". . . sinks the spirits of the church and king party everywhere, and ought to raise ours as much."[75] And it was, as his biographer Rutt recorded, not his sense of prudence, but his wife's, which enabled him to resist the pressures put upon him to attend the annual meeting of the Revolution Society on 4 November, at which the establishment of the new

---

[73] Ibid., "Part II," 38-9, and 39-41 for Priestley's election to the Convention in September. For Hurford Stone's correspondence from Paris, cf. T. S. 11/ 955/ 1793; and L. D. Woodward, *Hélène Maria Williams et ses Amis* (Paris, 1930, repr. Genève, 1977), 65ff.

[74] Priestley to Russell, 5 October 1792, *Works*, I.2.191-3. Hurford Stone's letter of 2 October is not among those transcribed in the series in T. S. 11/ 955/ 1793; but cf. his letter of 14 October, written in the "Camp before Verdun," in which he described himself as "domiciliated with one of the Generals who is an Englishman and who has distinguished himself much during the campaign." In November 1792 Stone presided in Paris over one violently revolutionary dinner of his English compatriots, the proceedings of which were reported in England. Cf. L. D. Woodward, *Hélène Maria Williams et ses Amis*, 65-75; Alger, "The British Colony in Paris," *E. H. R.*, 13 (1898): 672-4; Chaloner, "Joseph Priestley," 34.

[75] Priestley to Russell, 17 November 1792, *Works*, I.2.195; Priestley to Rev. Edwards, 16 October 1792, Priestley Collection, Dickinson Coll.

republic, and her victories abroad, were applauded with an extraordinary fervour.[76]

It was as a committed supporter of republican France, whose public letters to the Convention and its leaders added further to his notoriety[77] that Priestley, while apparently settling down in his place of retreat at Hackney, was nevertheless continually wrestling with the conflicting considerations which were to determine his emigration from England. This was, as he was increasingly forced to recognise, becoming a matter of urgent necessity. In the second part of his *Appeal to the Public*, published in November 1792, he defended the naturalisation of his son in France; he added that the conferring of French citizenship upon himself he considered "the greatest of honours"; and he spoke with understandable bitterness about his fellow countrymen: "As to myself, I cannot be supposed to feel much attachment to a country in which I have found neither protection nor redress."[78] Shunned by his fellow members of the Royal Society, his letters to Wilkinson in 1793 reveal how close an interest he continued to take in political affairs, how anxiously he watched as England and France went to war; and with what added apprehension he heard of the descent of France into anarchy. He was at this time much in the company of the Foxite Whigs. He gave an account of their discussions on the declaration of war, in which his views were clearly in accord with theirs: "That the French do not fear the war is evident enough, tho' it is as evident that they wished to avoid it, and were sincerely desirous of our friendship."[79]

While congratulating his brother-in-law that he was "out of the *Mania*, as you properly call it," he nevertheless took care to send him the pamphlets which appeared attacking the English ministry's stance on the war—praising in particular James Currie's pseudonymous "Jasper Wilson" *Letter*, and, also, Benjamin Vaughan's series of letters for the *Morning Chronicle*, under the pseudonym of "A Calm Observer," which appeared in pamphlet form in the summer of 1793. "If you should have any guess about the writer," Priestley wrote to Wilkinson, on sending him a copy of the latter, "I have a particular reason for desiring you would not give any intimation of it to *Mr. Vaughan.*" "Mr. Vaughan and others," he reported in July, "think that the ministry, as well as the nation, begin to

---

[76] Priestley, *Works*, I.2.367: "At the close of 1792, by the desire of some common friends, as well as from my own inclination," wrote Rutt, "I endeavoured to prevail on Dr. Priestley to take a very public part upon an interesting political occasion on which I had been appointed to preside. I allured him, I remember, among other inducements, by the example of Dr. Price, in 1789. At length, with his usual disregard of personal consequences, he freely assented to my proposal." It was Mrs. Priestley who convinced Rutt that "desire had outrun discretion." For an account of this dinner, cf. A. Goodwin, *The Friends of Liberty. The English Democratic Movement in the Age of the French Revolution* (Harvard Univ. Press, 1979), 246–7. For Priestley's attendance at the Society's dinner of the previous year, cf. "Revolutionary Philosopher, Part II," 36.

[77] Ibid., Appendix.

[78] Priestley, *Appeal to the Public*, Part II, *Works*, XIX.506 and note.

[79] Priestley to Withering, 15 April, 22 October 1793, *Scientific Correspondence*, 134 and n., 137. Priestley to Wilkinson, 18 February 1793, W. P. L.

be tired of the war, and would make peace if they knew how."[80] Priestley's relationship with Benjamin Vaughan was at this time particularly close: "There is no person, I believe, in England," he had written to Wilkinson on discussing his projected investments in France, "who is better acquainted with France, and French affairs, than he is; so that you may depend on any accounts that he may give you, and he wants no zeal to serve me, or my friends."[81] And throughout this period he clearly relied upon this inveterate enthusiast for the French cause as a valuable source of political information.

In March 1793, Priestley commented to Wilkinson on the passage of the Aliens Bill, forbidding all correspondence between England and France. It was, he thought, "very unreasonable and unnecessary," although it could not, he believed, "affect purchases already made in France." But, he added, it would "effectually prevent any person going to or coming from that country except with the approbation of the Court."[82] In this letter he told Wilkinson also of his son William's decision to leave France for America. Throughout April, with news from France now silenced, Priestley wrote to his friends of his concern for William, and of the increasingly distracted state of France. "We have no intercourse now with France," he wrote to Withering, "and whether my son William has been able to leave it and go to America I cannot learn. Indeed, the prospect is very melancholy. The conduct of the French has been such as their best friends cannot approve; but certainly the present combination against them . . . is as little to be justified."[83] "France, I fear, will long be in a lamentable state," he wrote to Wilkinson in May. "I have no fear on account of their foreign enemies, but their dissentions among themselves."[84] In August he could write more optimistically, reporting from a letter which certainly reached his radical circle in contravention of the Aliens Bill: "As I wish to give you all the information I can collect," he wrote to Wilkinson,

---

[80] Priestley to Wilkinson, 18 February, 6 April, 3, 15 July 1793, W. P. L.; and cf. Priestley to Wilkinson, 20 June 1793 (ibid.), sending him a copy of the *Morning Chronicle* with an account of the debate on Fox's motion for the re-establishment of peace with France, 17 June 1793 (*Parliamentary History*, XXX. 994–1024). For Benjamin Vaughan's two pamphlets published at this time—*Letters on the Subject of the Concert of Princes and the Dismemberment of France and Poland* (London, 1793); and *Comments on the proposed War with France, on the State of Parties, and on the New Act respecting Aliens* . . . (London, 1793), see Graham, *"Reform Politics,"* forthcoming; and Murray, *Benjamin Vaughan*, 300–30. And for Currie's pamphlet, cf. Graham, ibid.

[81] Priestley to Wilkinson, 4 October 1791, W. P. L.

[82] Priestley to Wilkinson, 19 March 1793, ibid.

[83] Priestley to Withering, 15 April 1793, *Scientific Correspondence*, 135; and cf. also Priestley to Wilkinson, 6 April, 16 May, 15 July 1793, W. P. L. For William Priestley's arrival in Philadelphia in the summer of 1793, see Rush to Mrs. Rush, 21 August 1793, L. H. Butterfield, ed., *Letters of Benjamin Rush* (Princeton Univ. Press, 1951), II.638: "A son of Dr. Priestley has just arrived in this city from France. He gives a most distressing account of the affairs of that country."

[84] Priestley to Wilkinson, 16 May 1793, W. P. L.; and cf. Priestley to Russell, March 1793, *Works*, I.2.196.

I shall observe that Mr. Stone, who is now at Paris, tells his brother, that there were proposals for peace from England in Paris on the 20th of May, but the revolution that followed prevented any thing being done. All affairs among themselves are likely to be settled very amicably, and they make no great account of their enemies.[85]

France, however, he had by now certainly, if reluctantly, recognised, could not offer at present, either for himself or his sons, a suitable asylum. "I perceive your resolution, and approve of it," he wrote in April to Russell on the intended emigration of the latter to America. "I take it for granted that I shall very soon be compelled to take the same measure. . . . Everything," he added, "indicates a beginning of troubles in Europe. I wish my friends, especially my young ones, safely out of it. As to such as myself it is of little consequence whether we go or stay."[86]

It was the plans and settlement of his sons, combined with the imperatives increasingly forced upon him by the political situation, which were to determine the manner and timing of Priestley's departure. For at much the same time that William Priestley left France for America, Priestley's other sons, Joseph and Harry, were coming to a similar determination to leave England. In his letter to Withering of 15 April Priestley had written that "great numbers are going to America, and among others all my sons," and of his intention to follow them.[87] In February in a letter to Adams he had already anticipated this move.

Such is the situation of this country, that I fear I shall be too troublesome in recommending to your notice Dissenters that are disposed to emigrate, and settle on your Continent. This letter will be delivered to you by two young men of good character, and fine spirit, the sons of Mr. G. Humphreys, a fellow sufferer with me in the Riot in Birmingham. Many others will also find it necessary to look out for an asylum either in France or with you; and at present America will (in) general be preferred to France. Two of my sons are in the number, and they will wait upon you in a few months, and if they get a settlement, I shall be happy to follow them.[88]

---

[85] Priestley to Wilkinson, 19 August 1793, W. P. L.; and cf. Priestley to J. Gough, 25 August 1793, *Works*, I.2.207. Cf. J. H. Stone to William Stone, 26 February 1793, T. S. 11/ 955/ 1793: "I will not hazard any more Letters on Politics by the Post Office." He nevertheless wrote frequently "by private hands" to his brother (same to same, 2 July 1793, ibid.), to others of their circle in London, and clearly frequently to Priestley: cf. Stone to William Stone, 22 November 1793, T. B. and T. J. Howell, eds., *A Complete Collection of State Trials* (London, 1809–1828), XXV.1211: "I refer you to a Note written to Dr. P. for what I have now scratched thro'"; and also same to same, 16 December 1793, ibid., XXV.1213: "Tell the Doctor that I have received his letter."

[86] Priestley to Russell, 30 April 1793, *Works*, I.2.199; and cf. Priestley to Lindsey, 5, 23 August 1793, ibid., I.2.206-7. Cf. also J. Hurford Stone to William Stone, 16 January 1794: "How is it that Dr. P. has received no letter from me, it would have opened his mind, which the detail I fear of the last six months has too much closed; tell him that I have all his fears and feelings, and yet I am more than satisfied, and in me it is a thousand times more meritorious": *State Trials*, XXV.1217; T. S. 11/ 955/ 1793.

[87] Priestley to Withering, 15 April 1793, *Scientific Correspondence*, 135.

[88] Priestley to Adams, 23 February 1793, Mass. Hist. Soc., Adams Papers, Reel 376; and cf. Adams's reply to Priestley, 12 May 1793, ibid., Reel 116: "Your Sons I shall be very glad to see and although it would give me great personal pleasure to see you in America, yet I cannot but think your removal would be a great loss to the philosophical and literary world."

To John Vaughan in Philadelphia, who had apparently repeated his offers of assistance, Priestley described his situation in a letter of some length. He emphasised his own continuing hesitation at leaving for such a distant asylum, and his undoubted preference for France, were her state not so unsettled, and the difficulties created by the war making a retreat there impossible. Joseph, he wrote, was "under a necessity" of leaving his situation in Manchester. He intended to go to America "to spend a year looking about him. Many others," added Priestley, "at least a hundred families, will also leave that neighbourhood."[89]

Joseph and Harry Priestley were to leave England in the company of many others, forced, with themselves, to leave a hostile social and political climate, in August 1793. From Manchester earlier in that year Joseph Priestley, Junior had described his own plight—of the determination of his partner, Ashworth, that he should "leave Manchester immediately"; of his own preference in many ways for France; and of the arguments of others: "Cooper Walker & all here prefer America which surprizes me," he wrote. "A more perfect government, the improved state of arts and sciences, the more general diffusion of knowledge, nearness to England & perhaps climate speak in favour of France, together with its being near the scene of action for many years to come." His parents, however, he believed, would be happier in America.[90] And it was perhaps the opinions of his father, as much as those of his political friends, which led young Priestley, as ardent an admirer of France as was his father, to make his plans to visit America shortly thereafter. Closest of all to him in this emigration, and apparently a constant companion on their travels in America, was one who from this time onwards was to be much associated with the Priestley family—at one time living in their household, certainly well acquainted with their fluctuating fortunes, and a steadfast admirer of Priestley's political views—the Manchester radical, Thomas Cooper. No account of Priestley's years in America would be complete without an understanding of the role of Thomas Cooper, and of his career in England, prior to his departure, with two of Priestley's sons, in the summer of 1793.

Thomas Cooper in the years 1790-3 had become one of England's most outspoken radical activists. A man of considerable erudition and scientific knowledge, a barrister by training, but his pursuits "chiefly literary and philosophical," he had had a sufficient knowledge of chemistry to start a bleaching manufactory in Manchester, where he was also a promi-

[89] Priestley to J. Vaughan, 6 February 1793, A. P. S., B. P. 931; and Appendix; and see also Priestley to C. Vaughan, 23 February 1793, Charles Vaughan Papers, Bowdoin Coll.

[90] J. Priestley, Jr., to Watt, Jr., 18 February, 16 March 1793, B. R. L. For Harry Priestley, cf. Priestley to Wilkinson, 6 April 1793, W. P. L.: "As he chuses an active employment, rather than a profession, and several of his companions at the College are going to occupy lands in America, he wishes to do the same and as Mr. Vaughan's sons will take him under their care for a time, and prepare him for either Agriculture, or commerce, I think I cannot place him better." [The College was Hackney, the Dissenting Academy at which Priestley taught gratis in 1791-4, and of which Thomas Belsham (1750-1829) was resident tutor and professor of divinity. (Memoirs, I.120-1. For Belsham, cf. D. N. B.)]

nent member of the Literary and Philosophical Society. In 1787 he had read before the Society his *Propositions Concerning the Foundation of Civil Government,* in which he set out in extreme terms the doctrine of the sovereignty of the people, and the right of popular resistance to oppression, and anticipated with something like enthusiasm the prospect of imminent revolution: "The structure of political oppression . . . begins now to totter: its day is far spent: the extension of knowledge has undermined its foundations, and I hope the day is not far distant when in Europe at least, one stone of the fabric will not be left upon another." In 1787–8 Thomas Cooper played a leading role in Manchester in the movement for the abolition of the slave trade. In 1790 he came even more strikingly into public prominence as a spokesman for the Dissenters. In March 1790, in the debate on the repeal of the Test and Corporation Acts, and subsequently, in the spring of 1792, after his visit to Paris with James Watt, Junior, the son of the great inventor, Thomas Cooper was denounced in the Commons by Burke for the propagation of his very extreme political principles. He was a founder member, with his close friend Thomas Walker, of the Manchester Constitutional Society, a frequent attender at meetings of radical societies in London, and an enthusiast for the writings of Paine. In the summer of 1792 he reasserted the principles of his *Propositions* in his *Reply to Mr. Burke's Invective.* His support for the revolution in France was, throughout the turmoil of that troubled year, unwavering.[91]

Both in his capacity as a radical activist and a scientist and man of letters, Thomas Cooper was well known to Priestley. In 1790 Priestley was instrumental in nominating Cooper for the Royal Society, and in his *Familiar Letters* of that year he deplored his rejection on what could only have been political grounds. "His knowledge of philosophy and chemistry far exceeds mine," he wrote.[92] In the spring of 1790, Priestley had conferred with Cooper in the campaign for repeal; and in 1791, when he was becoming increasingly anxious in his efforts to settle his son Joseph, it was to Cooper that he turned.[93] In the autumn of 1791 Thomas Cooper was among those members of the Manchester Literary and Philosophical Society who resigned when the Society refused to countenance an address of sympathy to Priestley. And, although there is little evidence of any intimate acquaintance between the two men—greatly separated as they were in age—nevertheless in the early autumn of 1792 Thomas Cooper was in London advising Priestley on the details of the

---

[91] T. Cooper, *Propositions respecting the Foundation of Civil Government* (1790); D. Malone, *The Public Life of Thomas Cooper (1783–1839)* (New Haven, Conn., 1926), 12–63; Seymour S. Cohen, "Two Refugee Chemists in the United States, 1794," *Proc. Am. Phil. Soc.,* 126.4 (1982): 304–10; Graham, *Reform Politics in England, 1789–99* (forthcoming).

[92] Priestley, *Familiar Letters, Works,* XIX.220–1; and cf. Priestley to Sir Joseph Banks, 25, 27 April 1790, *Scientific Correspondence,* 100–2; Cohen, "Two Refugee Chemists," 307–9.

[93] Priestley, *Works,* I.2.58; and Priestley to Lindsey, 9 January 1791, D. W. L.: passage omitted in Rutt: "I have written to Mr. Cooper to look out for a situation for my son at Manchester."

second part of his *Appeal to the Public*.[94] In this publication Priestley made another impassioned plea on Cooper's behalf, and went out of his way to endorse his very extreme political position. His "general abilities," wrote Priestley, "appear by his publications to be of the highest order"; and he specifically cited Cooper's contentious *Reply to Burke*. "He has given noble proofs of his public principles and his public spirit, and," he approvingly added, "he has been stigmatized by Mr. Burke."[95]

In the reaction of 1792-3, and the harassment of the Manchester reformers, which threatened to ruin several of them in their mercantile concerns, Thomas Cooper was to fall a victim. His bankruptcy was not announced until November 1793, but already his friends, among whom was young Priestley, were writing of the "terrible disaster that has befallen poor Cooper," and he himself had decided to "quit trade," and to leave "this most infernal and detestable kingdom" for America.[96] "All the Accts. of it I hear, confirm me in my inclination," he wrote. And he was by now also privy to the plans of the Priestleys, of Wilkinson's intention to divide his property in the French funds among Priestley's sons, and of his wish, too, were he "young as they are," to "be of the party."[97] In June Cooper wrote that he expected "to sail at the latter end of next month with young Priestley & *perhaps* with old Priestley."[98] Early in August, however, he wrote that "I & Joe Priestley & his brother set sail . . . from London to Philadelphia on the 15th Inst."; that his son was accompanying them; and "old Dr. Priestley and his wife in the spring." Thomas Walker, he confidently reported, was intending "to wind up his Concerns as quickly as possible to leave this detestable Country."[99] On 24 August, however, in the last of his letters to young Watt chronicling

---

[94] Malone, *Thomas Cooper*, 30-1; Priestley to Russell, 26 September 1792, B. M. Add. Mss., 44, 992.

[95] Priestley, *Appeal*, Part 2 (1792), *Works*, XIX.504-5; and Kramnick, "Joseph Priestley's Scientific Liberalism," 8-9: but the statement that Thomas Cooper became Priestley's "closest philosophical and religious follower" is, although on the first count surely correct, on the second, definitely not the case (cf. below, n. 111).

[96] Malone, 71; Priestley, Jr. to Watt, Jr., 14 April 1793, B. R. L.; and cf. Walker to Watt, Jr., 3 April 1793; Cooper to Watt, Jr., 10, 24 April 1793, ibid. Malone, who did not use Cooper's letters to young Watt in his account, doubted if the failure of Cooper's firm "played any direct part in causing his emigration"; and he commented also that Cooper "seems to have saved something from the wreckage." But cf. below, n. 100, for the extent to which Cooper's departure for America was to be dependent upon "the constant kindness" and financial assistance of his friends—among whom were both Priestley and his son Joseph.

[97] T. Cooper to Watt, Jr., 24 April 1793, B. R. L.; and cf. also T. Cooper to Wilkinson, 16 April 1794, W. P. L., where he refers to the letter he quoted to Watt, and hence his knowledge of Wilkinson's plans: "What I knew before from the Dr. who shewd me yr. letter."

[98] T. Cooper to Watt, Jr., 4 June 1793, B. R. L.; and cf. Priestley to Wilkinson, 20 June 1793, W. P. L.: "The ship that Joseph and Harry were to go in is arrived, but so much sooner than it was expected, that I rather think they cannot be ready to go in it, tho Joseph I perceive is very eager to be gone and Mr. Cooper wishes to go with them. By this time he is probably returned to Manchester." And cf. Priestley to Wilkinson, 15 July 1793, ibid.

[99] Cooper to Watt, Jr., 2 August 1793, B. R. L.; and cf. Priestley to Lindsey, 24 July, 5 August 1793, *Works*, I.204, 205. Cf. Priestley, Jr. to Watt, Jr., 14 April 1793, B. R. L., in which he also attempted to encourage young Watt "to think seriously about going to America, for depend upon it no good is to be done here. Walker & Cooper whose firmness & the

his disastrous last few months in England—his relinquishing "the whole" of his private property for the benefit of his creditors; his acceptance from a devoted circle of radical friends of funds to settle him in America—Thomas Cooper described the danger in which he believed himself also to be as a result of accompanying Walker for his hearing at Lancaster Assizes. He dared not, he wrote, board the ship at Gravesend:

> The reason why I am thus waiting is that my presence at Lancaster with Walker (whom I accompanied from London that I might see how the prosecutions agst. him wd. turn out) occasioned a writ or two to be taken out against me . . . Wher. there are any persons waiting on board the vessel to apprehend me, or wher. I shall get off safe from hence, is to me at this present writing, matter of anxious and disagreable Suspense.

He wrote also of his apprehension of war between England and America, and of its implications for England: "not," he added, "that the Ministry care for the *Commerce* of the Country: on the contrary I am persuaded there is a serious premeditated Intention to crush it, or at least most effectually to prune the wings of the Commercial and manufacturing Interests as being too provoking & dangerous to the aristocracy of the Country. And," he continued, "to this I am equally persuaded that if the combined Armies succeed agst. France they will certainly attempt to exterminate republicanism in America; & by Letters . . . from New York I find that the Americans see this in its true light."[100]

It was as refugees from a land inimical, as they believed, to their political as well as their economic well-being—Thomas Cooper bankrupt and under threat of prosecution, and young Joseph Priestley acutely aware of how little the future held for him in England—that "the voyagers," as Priestley called them, set sail in the late summer of 1793. Their ship, as Priestley wrote, was "crowded with passengers, as is every other ship that sails for *America*."[101] They were armed with letters of introduction,

---

Sanguininess (sic) of whose expectations you are acquainted with, will, I am sure, tell you the same. The country in my opinion is doomed to be enslaved." Thomas Walker, "the first manufacturer in Manchester," was a close friend of Cooper and Priestley. In 1793 he was given notice of prosecution by the government and brought to trial in April 1794 for alleged seditious activities. Cf. F. Knight, *The Strange Case of Thomas Walker* (London, 1957); Goodwin, *The Friends of Liberty.* For Priestley's concern for Walker in 1793, cf. his letter to Lindsey, 5 August 1793: "I cannot express how much I feel for him."

[100] Cooper to Watt, Jr., 4 June, 24 August 1793, B. R. L.: among the many members of the English radical fraternity whom Cooper in his letter of 24 August listed as helping him financially were "Old Priestley," Thomas Walker, William Russell, and also "young Priestley." Cf. also J. Priestley, Jr. to Watt, Jr., 14 April 1793, ibid. Malone, *Cooper,* 70, states that although one of the Manchester reformers was arrested in June 1793 and charged with having distributed the seditious paper signed "Sydney," which forcefully argued the case of the commercial interest against the war, and was widely known to have been written by Cooper, no indictment was preferred, and "so Cooper cannot be said to have been directly involved in any prosecution." Cooper's letter of 24 August 1793 appears to imply the contrary, however.

[101] Priestley to Lindsey, 7 September 1793, *Works,* I.2.209; Priestley to Wilkinson, 19 August 1793, W. P. L., and Cooper to Watt, Jr., 24 August 1793, B. R. L.: "The numbers emigrating from London Bristol & Liverpool are astonishing. The ship I go in has refused passage to near 100, and we are as many as that on board."

from Priestley himself, and from others, to many in America.[102] Their original intention, as Priestley described it, and also Thomas Cooper, was to travel to Kentucky: "at least they will *see* that country, and form their opinion there," wrote Priestley.[103] It was the declared intention of Joseph Priestley, Junior to visit Charles Vaughan in Boston at some stage on his travels. And that he and Cooper were pressed by both Jefferson and Madison to consider settling in Virginia seems, from a later letter of Jefferson's to Priestley, certain. With the active encouragement of John Vaughan, however, Cooper and young Priestley, after a lengthy trip along the Susquehanna, decided upon their ambitious plan of purchasing lands in a consortium.[104] And in December 1793, a freshly optimistic Thomas Cooper—albeit extremely apprehensive in the circumstances for his personal safety—left New York to return to England. "I will be at your house in February or March," he wrote to Samuel Rogers, in a letter which incidentally makes clear Priestley's very close connection with him at this time:

incog. like other great men. Mention this, with strong injunctions of secrecy, to Tuffin and Sharp . . . Russell, Priestley, and T. Walker (not R. Walker nor any other of my friends or my family) know of my intention. I hope to come over with a sufficient inducement for others to return with me.[105]

In England Priestley, increasingly oppressed by the political situation—unable, as Thomas Fysshe Palmer described it, "to sleep quietly in his

---

[102] Priestley to Adams, 20 August 1793, Mass. Hist. Soc., Adams Papers, Reel 376; Priestley to J. Williams, 23 August 1793, A. P. S., Misc. Mss. Colln.; Priestley to J. Gough, 25 August 1793, *Works*, I.2.207–8; Priestley to Thatcher, 21 August 1793, *P. M. H. S.*, Series 2, Vol. 3 (June, 1886): 16; Priestley to Rev. Dr. Abercrombie, 21 August 1793, *Scientific Correspondence*, 136; Priestley to Jedidiah Morse, 24 August 1793, Penn. Hist. Soc. Mss., Gratz Colln. Cf. also Malone, *Cooper*, 77 note, and below, Appendix, for the letters which Joseph Barnes wrote introducing Cooper and young Priestley to Jefferson.

[103] Priestley to J. Williams, 23 August 1793; and cf. Cooper to Watt, Jr., 4 June, 24 August 1793, B. R. L.: "We shall meet Toulmin at Philadelphia and we shall journey onward toward Kentucky"; "I shall therefore proceed straight to Kentucky of which the account printed & the account verbally related to me satisfy me completely." Rev. Harry Toulmin, dissenting minister, was the son of Rev. Joshua Toulmin, whose radical political opinions brought upon him much violence and obloquy, and who apparently considered emigrating to America with a large part of his congregation: *D. N. B.*; McLachlan, *Letters of Lindsey*, 121; and Brand Hollis to Adams, 18 February 1793, Mass. Hist. Soc., Adams Papers, Reel 115. Harry Toulmin did emigrate, and settled in Kentucky, where in 1796 he became Secretary of State: *Dictionary of American Biography* (hereinafter *D. A. B.*); McLachlan, *Letters of Lindsey*, 121; and Priestley to Willard, 10 April 1793, *P. M. H. S.*, Series 2. 43 (1910): 639–40.

[104] D. J. Jeremy, ed.,"Henry Wansey and his American Journal, 1794," *Mem. Am. Phil. Soc.*, 82 (Philadelphia, 1970): 77–9 and n. 86, 118–21; M. C. Park, "Joseph Priestley and the Problem of Pantisocracy," *Proceedings of the Delaware County Institute of Science*, 11.1 (Philadelphia, 1947); T. Cooper, *Some Information Respecting America* (London, 2nd edn.,1795), 85–6; 102ff; J. Priestley, Jr., to Charles Vaughan, 20 July, 22 November 1793, January 1794, Charles Vaughan Papers, Bowdoin Coll. For Jefferson's letter to Priestley, cf. below, n. 178.

[105] T. Cooper to Samuel Rogers, 14 December 1793, P. W. Clayden, *The Early Life of Samuel Rogers* (London, 1887), 286; and cf. J. Priestley, Jr. to J. Watt, Jr., 11 March 1794, B. R. L.; Cooper, *Some Information*, Preface, iii.

bed, owing to the unceasing persecution of the high church party"[106] –
was making definite preparations for departure. On 28 February he delivered to a crowded congregation his *Sermon* on the Fast Day, in the published Preface to which he dwelt at length on the harassment to which
he had been subjected, and, fearful himself now of prosecution, made
his extraordinary disclaimers of interest in politics. "Except in company,"
he declared, "I hardly ever think of the subject"– although, as he also
said, "I by no means disapprove of societies for political information,
such as are now everywhere discountenanced, and generally suppressed." In his *Sermon*, in his interpretation of the prophetic books, he
declared openly his belief in the imminence of revolutionary upheaval
throughout Europe: "May we not hence conclude it to be highly probable, that what has taken place in *France* will be done in other countries?"
On 30 March, to another crowded audience, Priestley delivered his Farewell Discourse to his congregation. "The time, I hope is approaching," he
said, "when all delusion will vanish; when men and things will be seen
in their true light; and the prevalence of truth will, no doubt, be attended
with an increase of general happiness."[107]

Priestley was still at this time in regular communication with Hurford
Stone in Paris, commenting to him apparently in terms of some disillusion on French affairs, although his name, it is clear, still carried
authority in Paris.[108] He was, above all, anxious to hear from his sons.[109]
And he was eager also, as is quite clear from his correspondence with
Wilkinson, to hear from Thomas Cooper. "The scheme of purchasing a
tract of land is Mr. Vaughan's," he wrote in January 1794,

who, as he lives in America, is the best judge of it, and as he himself embarks
as a principal, I am disposed to think well of it. Perhaps you may not dislike to
have some stake in that country, as well as in this. If so, I will desire Mr. Cooper,
who I understand is coming, to lay the scheme before you.

He wrote also of Russell's visit to him, and that he was "much interested in the scheme formed by Mr. Cooper and my son in America." "Mr.

---

[106] Fysshe Palmer to James Smiton, 20 July 1793, *State Trials*, XXIII.325. He had, wrote
Palmer, "been obliged more than once since he has been at Hackney to leave his house,
lest he should be burnt alive."

[107] Priestley, *The Present State of Europe compared with Ancient Prophecies; A Sermon
preached at the Gravel-Pit Meeting, in Hackney, February 28, 1794 . . . with a Preface, containing
the Reasons for the Author's leaving England, Works*, XV.519–552; Priestley, *The Use of Christianity, especially in difficult Times. A Sermon delivered at the Gravel-Pit Meeting, Hackney, March
30 1794, being the Author's Farewell Discourse to his Congregation, Works*, XV.553–569. Cf. "Revolutionary Philosopher, Part I," 46–8; "Part II," 42–4. For Priestley's distribution of these Sermons among his friends in England before his departure, cf. *Works*, I.2.229–30. For his
printing and distribution of them on his arrival in America, cf. below, n. 147.

[108] J. Hurford Stone to William Stone, 24 January 1794, *State Trials*, XXV.1219–21; T. S.
11/ 555/ 1793: and same to same, 27 February 1794, T. S. 11/ 555/ 1793: "Tell Eleanor that
I have received her Letter as also the Drs." For the authority which Priestley's name still
carried in France, cf. J. H. Stone to W. Stone (16 January 1794), *State Trials*, XXV.1215–17;
T. S. 11/ 955/ 1793; and same to same, 17 January 1794, *State Trials*, XXV.1217–19.

[109] Priestley to Wilkinson, 13 December 1793; 9, 25 January, 7 February, n.d. (February)
1794, W. P. L.

Cooper is not yet arrived," he wrote in a subsequent letter, "and some of his friends think he will not come at all. Indeed, I cannot help thinking it very hazardous for him."[110] And Thomas Cooper, when he finally did, at the end of March, reach the north of England was, as he promised, traveling "incog." He was, since he stayed in Lancaster for the trial of Thomas Walker on 2 April, unable, as he later wrote to Wilkinson, to see Priestley.[111] It was, however, now with the firm intention of establishing a settlement for his fellow refugees from England—"an asylum for my christian and unitarian friends," as he wrote; "a rallying point for the English, who were at that time emigrating to America in great numbers," as his son described it—that Priestley on hearing of the plans of his sons and Thomas Cooper, prepared to leave England.[112] On 7 April "the Dr." and his wife boarded the *Sansom* at Gravesend, bound for New York, with several other persons of "good property"—the majority in the cabin "aristocratically inclined," wrote Priestley, "but all in the steerage were zealous republicans"—and "accompanied by a vast number of emigrants of all descriptions."[113] "I do not pretend to leave this country, where I have lived so long, and so happily, without regret," he wrote in one of his last letters to Wilkinson from England: "But I consider it as *necessary*, and I hope the same good providence that has attended me hitherto will attend me still."[114] From Hackney, Sarah Vaughan, whose family had

---

[110] Priestley to Wilkinson, 25 January, 7 February 1794, n.d. (February) 1794: "If any body else come on the same business, I will endeavour that you shall see him." For Russell's departure with his family from Falmouth on 13 August 1794, cf. Jeyes, *The Russells of Birmingham*, 57-9.

[111] Cooper to Watt, Jr., 25 March 1794, B. R. L.; Cooper to Wilkinson, 16 April 1794, W. P. L. For the coolness which, shortly before his departure, Priestley suddenly felt for Thomas Cooper, on hearing that he had "become a determined unbeliever in Christianity," and on the implications of this for his joining the settlement in America, cf. his letter to Cooper, 6 April 1794, written the day before he left London, cited in D. J. Jeremy, ed., *Wansey's Journal*, 78-9, n. 82, and also below, n. 225. (The author is very grateful to Dr. Jeremy for providing her with a copy of this letter, the original of which is now missing.) It remains the case, however, as Malone pointed out (*Cooper*, 79), that Thomas Cooper was extremely close politically to the Priestleys throughout the emigration project, and this intimacy, in spite of their difference of religious belief, was to prove of lasting duration in America. The point is important, for Thomas Cooper's influence is frequently misinterpreted. His later cooperation in politics in America with Priestley should not be regarded as an aberration on Priestley's part (cf. Robbins, "Joseph Priestley in America," 74). For the effect which their difference of opinion in religious matters did have upon their relationship in their early years in America, however, cf. below, pp. 85-6.

[112] Priestley to Thomas Cooper, 6 April 1794, above, n. 111; Priestley, *Memoirs*, I.166; and cf. La Rochefoucauld Liancourt, *Travels through the United States of North America . . .* (London, 1799), 74; and R. W. Davis, *Dissent in Politics, 1780-1830. The Political Life of William Smith, M.P.* (London, 1971), 75. In the *Memoirs* young Priestley wrote—in contrast to his father—that "the scheme of settlement was not confined to any particular class or character of men, religious, or political." (*Memoirs*, I.166).

[113] Priestley to Lindsey, 6 June 1794, Priestley, *Works*, I.2.244; ibid., I.2.228 and note; and cf. also Priestley to Lindsey, 7, 9, 11 April 1794, ibid., I.2.229-31: "The only woman cabin passenger is come from France," Priestley wrote on 11 April, "knows our friends there, and seems well acquainted with the politics of the country."

[114] Priestley to Wilkinson, 7 February 1794, W. P. L.; and cf. also same to same (n.d., February 1794): "to abandon an advantageous, and agreeable situation for such an uncertainty, so late in life, is sometimes rather painful, but it is absolutely necessary."

always so unreservedly provided hospitality, solace, and support for Priestley, wrote to her son Charles in Boston of their sense of loss, and of Priestley's sufferings: "He has been persecuted with great severity. . . . I need not say anything to enforce your attentions to them. He will be an honour to any state that fosters such a Man."[115]

The sense of urgency surrounding Priestley's departure–"all my friends," he wrote, "advise me to go as soon as I can"[116]–was the result of the political crisis which many had long been expecting in England, and which was brought to a head in the early months of 1794. It arose in large part from the continuing activities of a vocal minority of English and Scottish reformers, their advocacy of a Convention in England, and their savage suppression by the Scottish authorities, with the explicit approval of the English ministry. If Priestley and his immediate circle were not themselves involved in these particular activities, nevertheless, their intimacy with many who were, their concern for their fate, and their certain approval of their aims–together with their continuing sympathy with France in general, and with John Hurford Stone in Paris in particular–was leading them also into dangerous waters. Priestley himself, in two little-quoted letters to Wilkinson, while making clear his alarm at the progress of political events, did not hesitate to make plain, beyond all doubt, his sympathies with the extreme views of the English and Scottish reformers, and his clear understanding of the continuity in their propagandists' writings of all that had been so forcefully put by Paine in 1791 and 1792. "Many persons," he wrote on 9 January 1794,

begin to apprehension (sic) an invasion from the French; and that the French really *intend* it I have no doubt I have had further evidence of it this very hour, in an account of a letter from Paris.

All these things make the present time look very serious. There are two excellent pamphlets just published, one called *Peace and Reform against War and Corruption* and the other *A Convention the only way* (sic: *means*) *to prevent ruin*; and I would have sent them, but they are too large to be sent in covers. They are as bold as Mr. Paine's writings, and more correctly written.

And he recommended to Wilkinson also "*Mr. Muir's Defence* in his Trial, as printed for Robertson," as "most excellent." The fate of Muir and Fysshe Palmer, whom he visited in Newgate after their farcical trials and before their barbarous deportation to Botany Bay, made a deep impression upon Priestley, and was, as he freely acknowledged, to be one of the factors determining his precipitate departure.[117] In a subsequent letter to Wilkinson, he wrote of the other considerations which were

[115] Sarah Vaughan to C. Vaughan, 25 March 1794, Charles Vaughan Papers, Bowdoin Coll.

[116] Priestley to Wilkinson, n.d. (February 1794), W. P. L.

[117] Priestley to Wilkinson, 9 January 1794, ibid.; and cf. "Revolutionary Philosopher, Part I," 46, and nn. 16, 17; Goodwin, *The Friends of Liberty*, 287–9. For Priestley's contribution to the subscription for Muir and Fysshe Palmer, and for Rev. Winterbotham, a Dissenting Minister whose successful prosecution by the Ministry also affected him deeply, cf. his letter to Lindsey, 6 December 1795, *Works*, I.2.325. For the two pamphlets, by Daniel Stuart and Joseph Gerrald, cf. Graham, *Reform Politics*.

undoubtedly affecting his state of mind, in particular the continuing belief of some in his circle that the French would soon attempt to invade England: "Others, however, and among them Mr. B. Vaughan, think it will come to nothing." But, he added, in a revealing passage,

Violent as the ministry are, I much fear there is as violent a spirit of opposition rising up in many of the *lower orders*, which threatens something even worse, while persons of property and moderation, will not be heard, but will be in danger of being crushed by both. The calmest of my friends are more alarmed than ever.[118]

In 1791, after the rioting against him, and while he was composing his intransigent and in many ways extreme *Appeal to the Public*, Priestley had not been in any doubt as to the course of action he should take: "We must not despair, or discover any timidity," he had written. "I rather fear going into the opposite extreme, which, however, I think is the better of the two."[119] In 1792 he had unreservedly welcomed the revolutionary writings of Paine and Barlow. And in 1794 he was still, at least in private, prepared to countenance the views of Gerrald, whose call for a Convention, and sweeping change in the constitution of England, was at the core of the reformers' activities. The prospect of imminent revolution, however — although he might in private declare it to be necessary, and in his writings also state that it must not only be allowed for but expected, as a valuable catalyst for change — was not one which Priestley as a principal participant could relish. His by now very understandable fears of the inevitable violence and confusion inherent in such a situation undoubtedly hastened his departure for America. That, nevertheless, he was closer to the scene of action than is usually appreciated — and that this too, was a factor in determining his hasty departure — can be inferred both from his correspondence with Wilkinson, and from a deposition which he was to make for John Vaughan, shortly after his arrival in America. For in this deposition Priestley was to affirm that William Stone had brought to his house in Hackney an account of a letter from his brother in France, asking him to canvas his friends in England on the state of opinion there should the French invade: that this was the letter he described in his letter to Wilkinson of 9 January seems entirely probable.

John Hurford Stone was by 1794 part of a larger conspiracy, of which the English ministry were by now well aware, and his effective implication of his political acquaintances in London in this was to have dire consequences. As a result of William Stone's response to his brother's enquiries — which he showed not only to Priestley and Benjamin Vaughan, but to others of their circle — he was arrested on 3 May on a charge of high treason, interrogated by the Privy Council, and held in Newgate

---

[118] Priestley to Wilkinson, 25 January 1794, W. P. L.
[119] Priestley to Lindsey, 30 August 1791, *Works*, I.2.151.

awaiting trial for two years.[120] Benjamin Vaughan, who composed for Stone a memorandum which in certain passages could certainly be regarded as compromising, was, with several others, requested to appear before the Privy Council on 8 May. His interrogation "lasted until near six o'clock"; and he was to lose no time in fleeing to France, and subsequently Switzerland. There he ignored the pleas of his patron Lansdowne, that he either "return quietly back," or "write a Letter stating his motives, and all that he did do and all that he did not do, and leave it to his friends here to publish it or not as may be most for his credit and theirs." Benjamin Vaughan, whose writings and "general conduct," as his father was to declare, had rendered him "the most obnoxious man of any in the Kingdom to the Ministry," and who had as a result "nothing to expect but the most severe persecution," remained in France and Switzerland until his eventual arrival in America in the summer of 1797. Even there, he wished to live as anonymously as possible.[121]

The arrest of William Stone, and the interrogation of so many of his close political acquaintance before the Privy Council was followed shortly afterwards by a further wave of arrests and interrogations of other prominent reformers in England. In London Thomas Cooper dined with Horne Tooke the night before the arrest of the latter. "You will see by the papers of today how prettily we are going on," he wrote to young Watt: "I rejoice that I have not long to stay (I *hope*) in this rascally Country. I fancy the ministers will not proceed further back than the Declaration of War between France & this Country, otherwise you & I wd. have the honour of an Examination before Mr. Pitt & Mr. Secretary Dundas . . . neither you nor I are in much Danger, in *my* opinion," he added.[122] But Cooper himself, while seeing his friends in Manchester, Birmingham, and London, and writing and publishing his pamphlet, *Some Information Respecting America*, in response to the many enquiries with which he was beset "respecting the state of Society, the means of living, and the inducements to settle upon that continent," now lost little time in leaving England for the town which he was hoping would be called "ASYLUM." His ship was due to sail from Liverpool on 10 August—"jour memorable,"

---

[120] L. D. Williams, *Hélène Maria Williams*, 101–112; Goodwin, *The Friends of Liberty*, 322–4; M. Elliott, *Partners in Revolution. The United Irishmen and France* (Yale Univ. Press, 1982); Graham, *Reform Politics*, forthcoming.

[121] Lansdowne to S. M. Vaughan, 5 June 1794; S. Vaughan to B. Vaughan, 13 August 1795, A. P. S., Vaughan Papers, B. V. 46 p. For Vaughan's memorandum, cf. *State Trials*, XXV.1236–9; and for a report of his interrogation before the Privy Council, which Priestley must have read on his arrival in Philadelphia, cf. *Philadelphia General Advertiser*, 3 July 1794; and below, n. 173. The account of this episode in Vaughan's career in Murray, *Benjamin Vaughan*, 336ff., is hampered by the fact that he could not locate the compromising memorandum (see especially, 339, note). Cf. however, M. V. Marvin, *Benjamin Vaughan* (Privately printed, 1979), 30–4. For Vaughan's arrival in France, his sojourn in Switzerland, and return to France, cf. ibid., 35–54; Murray, *Vaughan*, 343ff., and below, n. 272.

[122] T. Cooper to J. Watt, Jr., 19 May 1794, B. R. L.

he characteristically wrote. He wished, he said, "most sincerely my friends were ready for flight. This is not a Country to stay in."[123]

Priestley departed for America shortly before these dramatic developments, but already embodying the hopes—and now, also the financial investment—of many in England who were increasingly fearful as to the turn that events might be about to take. He left, as seems indubitably clear (for all the unique unpopularity which his outspokenness in matters of religious belief had brought upon him, and for all the over-riding importance which the propagation of these beliefs had for him) as a political exile, in the company of others who were similarly convinced of the impossibility in the present political climate of making a livelihood in England—and with more than one of his close acquaintance either imprisoned, deported, or under the shadow of prosecution. On his arrival in America, expecting still a safe haven and kindred spirits in the projected settlement on the Susquehanna, Priestley was to allow himself a full expression of his republicanism, of his continuing hostility to the government of England, and of his loyalty to the cause of France. He did, however, consistently avow his aversion to any involvement in the politics of America. That this was to be almost from the outset effectively impossible—that by his very presence, his reputation (and that of those with whom he was so closely associated), and by his actions, he was to remain a political figure who could not long remain unnoticed, it is in large part the purpose of this study to demonstrate.

Priestley's dilemma in America was that of the quondam revolutionary propagandist seeking asylum in the land whose freedom, "at some risk" to himself, as he wrote, he had helped to secure;[124] and whose interests had remained very close to his heart. He had, nevertheless, as he wrote to John Adams shortly after his arrival, "made it a rule to take no part whatever in the politics of a country in which I am a stranger, and in which I only wish to live undisturbed as such."[125] In spite of such declarations, however, Priestley, both on his arrival and subsequently, could not resist, at a time of great political tension in America, pronouncing upon the issues at stake. And finding himself besieged, he could not fail, when sufficiently provoked, to reply in kind. His plight, and the harassment which even in America he endured, underscores not only his unwavering adherence to his original political vision; it reveals also the revolutionary nature of this decade—its extremes of opinion and reaction. "When the times are so dark and serious with respect to nations," Priestley wrote in characteristic vein to Lindsey in 1798, "how

---

[123] T. Cooper, *Some Information Respecting America*, Preface, iii; Cooper to Wilkinson, 16 April 1794, W. P. L.; Cooper to Watt, Jr., 25 March, 13 April, 15 June, 9, 18 July, 17 August 1794, B. R. L. Cf. Malone, 79–80, and note.

[124] Priestley to Thatcher, 23 January 1800, *P. M. H. S.*, Series 2, Vol. 3 (June 1886): 31; and cf. also same to same, 26 July 1798, 7 January 1799, ibid., 24, 25.

[125] Priestley to Adams, 13 November 1794, Mass. Hist. Soc., Adams Papers, Reel 378.

can individuals expect to escape troubles?"[126] For if his inability even in America to escape controversy was certainly due in part to his own character—and, as many believed, to the unfortunate influence of Thomas Cooper—it was due in part also to the extremes of politicisation from which at this time America was far from immune. Priestley's continuing advocacy of the cause of France throughout this period, his unwavering hostility to the English government, and his leading role in a very general emigration of English—many of them of similar political opinions—are important factors to bear in mind in considering his years in America. But, as a corollary of this, the extent to which a reaction to the extremes of the republican ideal which the Revolution in France was seen to represent, and a movement of opinion in America not averse to altering the emphasis of her already more conservative Constitution was gathering force throughout the 1790s, must be fully appreciated. The change, as Priestley so frankly (but not without disingenuousness) remarked, in his controversial *Letters to the Inhabitants of Northumberland* of 1799, "is not in me, but in the people here."[127]   -

In 1792 John Adams, now the Vice-President of America, although not even at that time in such sympathy with the French as was Priestley, had nevertheless written to condole with him on his "sufferings in the cause of liberty." "Inquisitions and Despotisms are not alone in persecuting Philosophers," wrote Adams. "The people themselves we see, are capable of persecuting a Priestly, as another people formerly persecuted a Socrates."[128] In 1793, however, in reply to one of Priestley's several letters to him of that year, Adams wrote of his own conviction of the value of "the monarchical part" of the English Constitution: the "sublime and beautiful fabrick of the English Constitution in three Branches," was, he believed, essential for the preservation of the liberties of Europe.[129] Such thinking, increasingly applied to the troubled and divided politics of America, was to lead Adams, under the stress of the threat of French expansionism, to enact, as President, legislation which was thought in its potential for repression to exceed even that passed by Pitt's ministry in England. And in 1798 he wrote, in defence of his actions, that "we have had too many French philosophers already, and I really begin to think, or rather to suspect, that learned academies, not under the immediate inspection and control of government, have disorganized the world, and are incompatible with social order."[130] From his Secretary

---

[126] Priestley to Lindsey, 6 September 1798, *Works*, I.2.407.

[127] Priestley, *Letters to the Inhabitants of Northumberland and its Neighbourhood, on Subjects interesting to the Author, and to them* (2nd edn., Philadelphia, 1801), *Works*, XXV.119; and cf. Priestley to Thatcher, 7 January 1799: "The change, dear sir, is in you." *P. M. H. S.*, Series 2, Vol. 3 (June 1886): 26.

[128] Adams to Priestley, 19 February 1792, Mass. Hist. Soc., Adams Papers, Reel 115, and Appendix.

[129] Adams to Priestley, 12 May 1793, ibid., Reel 116.

[130] Adams to Pickering, 16 September 1798, C. F. Adams, ed., *The Works of John Adams*, VIII. 596.

of State came the advice on those two Englishmen who had, in spite of the failure of their projected settlement, remained in the town of Northumberland, Pennsylvania — Priestley, and Thomas Cooper. "Those who are desirous of maintaining our internal tranquillity," wrote Timothy Pickering, "must wish them both removed from the United States."[131]

---

[131] Pickering to Adams, 1 August 1799, ibid., IX.5–6.

THE AMERICAN POLITICAL SCENE IN 1794
AND THE ARRIVAL OF PRIESTLEY

In the pamphlet which was to encourage the emigration of so many Englishmen, philosophers and others, to settle in America during these years, Thomas Cooper in 1794 dwelt little on the already growing political strife in that country. "There is little fault to find with the government of America," he wrote, "either in principle or in practice . . . the present irritation of men's minds in Great Britain, and the discordant state of society on political accounts is not known there. The government is the government *of* the people, and *for* the people." There were, he acknowledged, two distinct parties in America: the Federalists, who leaned towards "an extension rather than a limitation of the powers of the legislative and executive government," who inclined "rather . . . to British than to French politics"; and to the introduction and extension of "the funding, the manufacturing, and the commercial systems." And there were the "Anti-federalists," so called "not because they are adverse to a federal government, or wish like the French for a republic, *one and indivisible*, but in contradistinction rather to the denomination of the other class." They had been, at the time of the framing of the Constitution, and they still were, "hostile to the extensive powers given to government," and they were averse to much of the present administration's policies. They inclined rather "to the French theory, though not to the French practice of politics"; and they were increasingly hostile to the arrogance of the English government. The Federalists, wrote Cooper, were "the 'ins', and the Anti- federalists the '*outs*'" of American politics. But in such matters, he wrote, in a sentence which in a very few years he certainly lived to regret, "we are more moderate than you are."[132]

Cooper's determination in 1794 to emphasize the inherent stability of America's democratic institutions, and to praise the prevailing tone of her politics, was echoed by Priestley in his early pronouncements on her government on his arrival there. But that he, too, realised that unanimity was not universal in America on the most pressing political issue of the day is clear from one of his letters written before his departure. "That many viewed" the French Revolution "in an unfavourable light, with you I have no doubt," he wrote. "But that a revolution so nearly resembling your own should not be thought a joyful event by the Americans in general, I could not be brought to believe." His correspondent had, he wrote, made him "quite easy on the subject," and enabled him to satisfy his

---

[132] T. Cooper, *Some Information respecting America*, 52; 67–9.

friends.[133] Priestley's own belief in the essential interdependence of the two revolutions in France and America, which no minor differences as to forms could dispel, had been expressed in all his contributions to the revolutionary debate in England. They heralded, he declared in the last chapter of his *Letters to Burke*, "a totally new, a most wonderful and important aera in the history of mankind." In the eagerly awaited *Discourse* which he delivered to the young Dissenters at Hackney in April 1791, he numbered the heroes of France and America together. And in his *Political Dialogue* in the summer of that year, written shortly before the riots and the destruction of his house, he stated that "in America and France . . . we have examples of two entirely new constitutions of government that deserve particular notice, as differing from any that the world has seen before."[134] In the *Dialogue* he entered at some length into a disquisition on the inherent value of the French experiment of a unicameral legislative assembly. "A national assembly thus constituted and frequently changed, could not have any other object in their consultations than the interest of the whole community," he wrote. Its measures would of course favour "*the greater number*, which ought to be the object of every government." "In every state, as in every single person," Priestley added,

there ought to be but one will, and no important business should be prevented from proceeding, by any opposite will. If there be two wills, and they can effectually counteract each other, it is no longer one state and one government, but two states and two governments, which though they may agree to act in concert, may likewise act separately.

And in this endorsement of the French republican experiment, he had even gone so far—while praising America for her timely abolition of all hereditary distinctions—as to question her constitutional system of a balance of powers, and to doubt the wisdom or necessity of the absolute negative which the senate could exercise "on all the resolutions of the house of representatives."

There does not . . . seem to be any necessity for different powers in the same state, each having an absolute negative on the proceedings of the other, in order to secure the most deliberate discussion of every public measure. If the majority of any people understand their own interest, there can be no good reason why they should not have the power of promoting it, and that with as little obstruction and delay as possible.[135]

Priestley's position in the debate engendered by the constitutional innovations of the French in implementing their Revolution, and his persistence in defending them, in public at least, at all times, had made him a target of abuse for Burke and all those in England who feared the effects

[133] Priestley to J. Gough, 24 August 1793, *Works*, I.2.207.

[134] Priestley, *Letters to . . . Burke*, *Works*, XXII.236ff; Priestley, *The Proper Objects of Education in the Present State of the World, represented in a Discourse, delivered . . . to the Supporters of the New College at Hackney* (London, 1791), *Works*, XV.422; Priestley, *A Political Dialogue on the General Principles of Government* (London, 1791), ibid., XXV.83.

[135] Priestley, *Political Dialogue*, 88–96.

of the French experiment. But it was to one in America who in his out-spoken opposition to the Revolution in France had taken up a position akin to that of Burke, that Priestley confided one of his few recorded remarks expressing some doubt about the government of France. "I cannot say but I now think more favourably of a pure republic than I have done," he wrote to John Adams in December 1792. "A comparison between the American and French governments some years hence will enable us to form a better judgment than we can at present."[136] His defence of France, however, and his attacks upon Burke, had gained him much praise from Francophile opinion in America. His reply to Burke was, thought Benjamin Rush, as "masterly" a performance as that of Paine, "although they possess different species of merit. Paine destroys error by successive flashes of lightning. Priestley wears it away by successive strokes of electricity." Jefferson, too, held Priestley's writings in as high esteem as those of the author of the *Rights of Man*. "The Revolution of France does not astonish me so much as the Revolution of Mr. Burke," he wrote to Benjamin Vaughan in the summer of 1791. "I wish I could believe the latter proceeded from as pure motives as the former. But what demonstration could scarcely have established before, less than the hints of Dr. Priestly and Mr. Paine establish firmly now."[137]

By the time Jefferson wrote this letter, the debate on the Revolution in France had burst out with great vehemence in America. "We have some names of note here who have apostatised from the true faith," Jefferson informed Benjamin Vaughan; and undoubtedly, as his later statements reveal, one of those whom he had in mind was John Adams. For Adams, in a series of Letters entitled *Discourses on Davila*, published in the *Gazette of the United States* from April 1790 until the following year, had given vent to the doubts which he had at a very early stage expressed about the innovative philosophy underlying the Revolution in France, and its implementation in the unicameral, sovereign assembly which Paine, Priestley, and so many other advocates of the French at this time defended. In so doing, Adams mocked also the prevailing belief in human perfectibility; and he came perilously close (as he had done in his earlier *Defence of the Constitutions of Government of the United States of America*, in which he had stressed the need for a properly balanced government in order to take account of the human need "for consideration, congratulation, and distinction") to a defence of a hereditary system of government. The governments of Europe, which had possessed such superiority in so many spheres, had, he acknowledged, been deficient in the representation of the people in government. The people should, if their interests

---

[136] Priestley to Adams, 20 December 1792, Mass. Hist. Soc., Adams Papers, Reel 375, and Appendix.

[137] Rush to Belknap, 6 June 1791, L. H. Butterfield, ed., *Letters of Benjamin Rush*, I.582–4; Jefferson to Benjamin Vaughan, 11 May 1791, J. P. Boyd and R. W. Lester, eds., *The Papers of Thomas Jefferson* (Princeton Univ. Press, 1982), XX.391; and cf. also Madison to Jefferson, 1 May 1791, R. A. Rutland, T. A. Mason et al., eds., *The Papers of James Madison* (Univ. Press of Virginia, 1983), XIV.15; and cf. below, n. 161.

were "honestly and prudently conducted by those who have their confidence . . . most infallibly obtain a share in every legislature."

But if the common people are advised to aim at collecting the whole sovereignty in single national assemblies, as they are by the Duke de la *Rochefoucauld* and the Marquis of *Condorcet;* or at the abolition of the regal executive authority or at a division of the executive power . . . they will fail of their desired liberty, as certainly as emulation and rivalry are founded in human nature, and inseparable from civil affairs . . . it is a sacred truth . . . that a sovereignty in a single assembly must necessarily, and will certainly be exercised by a majority, as tyrannically as any sovereignty was ever exercised by kings or nobles.

Adams further "respectfully insinuated: Whether equal laws, the result only of a balanced government, can ever be obtained and preserved without some signs or other of distinction and degree?" America, he was writing privately, if regretfully, to Benjamin Rush at this time, would, he was sure, eventually have to resort to two hereditary branches of government, "as an Asylum against Discord, Seditions and Civil War. . . . Our ship must ultimately land on that shore or be cast away."[138]

In May 1791 the first American edition of the *Rights of Man* was published in Philadelphia, and to it the printer appended a private note of Jefferson's, expressing his pleasure that it was to be reprinted in America, "and that something is at length to be publicly said against the political heresies which have sprung up among us." These, as Jefferson freely admitted to Madison in the ensuing furore, were "certainly the doctrines of Davila," although, as he added, "I tell the writer freely that he is a heretic, but certainly never meant to step into a public newspaper with that in my mouth." Adams, as Jefferson foresaw, was "displeased" with the imputation. In a subsequent letter to Jefferson he protested against the constructions which he claimed were wrongly put upon his writings, which had led to "floods and Whirlwinds of tempestuous Abuse, unexampled in the History of this Country," falling upon him as a result. His cause was taken up by his son, and the Publicola letters of John Quincy Adams continued the controversy over the proper implementation of republican principles throughout the summer of 1791.[139]

It was this debate, over the widely varying emphases and differing interpretations of republicanism—given fresh point with every new threat to the survival of the French experiment, and each successive stage of the descent of France into anarchy and dictatorship—that served as the background to the massive task facing Washington's Administration: the securing of the stability of America, which throughout the 1780s—

---

[138] Jefferson to Benjamin Vaughan, 11 May 1791; Adams, *Discourses on Davila; A Series of Papers on Political History by an American Citizen, Works,* VI.242-3; 252; 270-81; Z. Haraszti, *John Adams and the Prophets of Progress,* 20; 38ff; 165-6; J. R. Howe, Jr., *The Changing Political Thought of John Adams* (Princeton Univ. Press, 1966), 172-188 and especially n. 114.

[139] *Jefferson Papers,* XX.268-313: "The Contest of Burke and Paine in America"; M. D. Peterson, *Adams and Jefferson. A Revolutionary Dialogue* (Univ. of Georgia Press, 1976), 56-61; L. Banning, *The Jeffersonian Persuasion. Evolution of a Party Ideology* (Cornell Univ. Press, 1978), 154-9, 210-11; Cappon, ed., *Adams-Jefferson Letters,* I.245-52.

and even after the framing of the Constitution of 1787–remained in doubt. To the Administration led by Washington, Jefferson, and Hamilton was given the enormous task of framing a financial system for the federal government; determining the course which her commercial and economic policy was to take; taming the vast resources of her interior, and establishing her frontiers. And in this process, for Jefferson and the supporters of France, the successful issue of the French Revolution was vital for America–for, as Jefferson wrote, "I feel that the permanence of our own leans in some degree on that; and that a failure there would be a powerful argument to prove there must be a failure here." "The establishment and success" of the French government, he wrote, was "necessary to stay up our own and to prevent it from falling back to that kind of Halfway house, the English constitution." For the Federalists, however, and all those who, with Adams, had become increasingly convinced of the need for the strengthening of the executive, and greater checks upon the unbounded power of the people, the trials of France, and the machinations of her partisans in America, were but fresh proof of the dubious permanence of truly republican government, of the threat to the stability of America which France represented, and an occasion for further ambiguous encomiums on the advantages of the mixed constitution of England. "I said, that I was *affectionately* attached to the Republican theory," wrote Hamilton in May 1792: ". . . & I add that I have strong hopes of the success of that theory; but in candor I ought also to add that I am far from being without doubts. I consider its success as yet a problem." The enemies of the securing of "that *stability* and *order* in Government which are essential to public strength & private security and happiness," were, he wrote, "the Spirit of faction and anarchy."[140]

Increasingly, however, it was Hamilton's financial policies, which seemed to many a tame imitation of and essentially dependent upon those of England, which were to become a cause of friction between him and Jefferson, and of massive unrest in the country. Increasingly, too, the threat of French expansionism, and the policy pursued by the French republic of regarding America as a legitimate sphere of influence–a policy which became unmistakably clear with the visit to America of the French envoy, Genet, in 1793–threatened to divide the Administration. The follies and indiscretions of Genet alienated even Jefferson, but his partisanship for France was reflected in the Democratic Societies which were developing in considerable numbers along the eastern seaboard, their members toasting and celebrating the victories of the French, and violently opposed to many of the measures of the Administration. In the summer of 1794 the discontent against the government broke out into insurrection in Pennsylvania, in a revolt, backed almost certainly by

[140] Jefferson to Rutledge, 25 August 1791; Jefferson to George Mason, 4 February 1791, *Jefferson Papers*, XXI.75; XIX.241; Hamilton to Carrington, 26 May 1792, H. C. Syrett and J. E. Cooke, eds., *The Papers of Alexander Hamilton* (Columbia Univ. Press, 1966), XI.444; S. Elkins and E. McKitrick, *The Age of Federalism. The Early American Republic, 1788–1800* (O. U. P., 1993), 77ff.

some of the Democratic Societies, against Hamilton's imposition of an excise. The Whiskey Rebellion, as it was subsequently to be called, was the first occasion on which federal troops were called in to suppress a state disturbance. And with Washington's denunciation of the Democratic Societies as fomentors of the insurrection—"one of the extraordinary acts of boldness of which we have seen so many from the faction of monocrats," as Jefferson described it—and the simultaneous despatch of an envoy to England to negotiate an alliance with that country, the republican opposition to the Administration reached fresh heights.[141]

It was to this land of supposed political harmony, but in fact of increasingly partisan political discord, doctrinal dispute, and simmering internal discontent, that Priestley, with so many other English emigrants, arrived in the summer of 1794. In the prevailing political atmosphere, it was undoubtedly his renown as a propagandist in the cause of France, as much as his strenuous and long-lasting partisanship for the liberties of America, that influenced the more extreme voices in the general and triumphal welcome which he was accorded when, on 4 June 1794, after "a passage of eight weeks and a day," his ship docked at the Battery in New York. There his reception was, as he wrote to Lindsey, "too flattering, no form of respect being omitted." Among the many marks of attention which he received was a letter from John Adams, who on 3 June had left New York for Boston, but who had, on receipt of a letter which Priestley had asked to be delivered to him, left for him a message, that "he should be glad to see him at Boston, which he . . . thought better calculated for him than any other part of America, and that he would find himself very well received if he should be inclined to settle there."[142] In New York itself, Priestley was "received with a fervour of affection, which no king ever yet received, much less deserved," ran one account. "The town had been some time expecting his arrival," wrote Henry Wansey, a fellow Englishman and wealthy West of England clothier who watched with interest the honours showered upon Priestley, "and several societies intended shewing him particular honor." His arrival

---

[141] P. S. Foner, *The Democratic-Republican Societies 1790-1800* (Greenwood Press, 1976); E. P. Link, *The Democratic-Republican Societies, 1790-1800* (Columbia Univ. Press, 1942); G. Chinard, *Thomas Jefferson, The Apostle of Americanism* (2nd ed., repr. 1975); M. D. Peterson, *Thomas Jefferson and the New Nation* (O. U. P., 1970), 458ff; Banning, *The Jeffersonian Persuasion*, 181ff; Elkins and McKitrick, *The Age of Federalism*, 354-65, 455-88; Jefferson to Madison, 28 December 1794, P. L. Ford, ed., *The Works of Thomas Jefferson* (New York, 1904-5), VIII.156-9. For the Whiskey Insurrection, cf. W. Miller, "The Democratic Societies and the Whiskey Insurrection," *P. M. H. B.*, 62 (1938): 324-49; and T. P. Slaughter, *The Whiskey Rebellion. Frontier Epilogue to the American Revolution* (O. U. P., 1986).

[142] Priestley to J. Vaughan, 3 June 1794, A. P. S., Priestley Papers, B. P. 931; Priestley to Lindsey, 6 June 1794, *Works*, I.2.244-6; D. J. Jeremy, ed., "Henry Wansey and his American Journal," 89. It was Wansey whom Priestley had asked to deliver the letter to Adams. For Priestley's thoughts of settling in the neighbourhood of Boston, which he abandoned after hearing of the proposed settlement of his sons, cf. Priestley to Charles Vaughan, 23 February 1793; Sarah Vaughan to C. Vaughan, 25 March 1794, Charles Vaughan Papers, Bowdoin Coll.; and Priestley to the President of Harvard, Joseph Willard, 10 April 1793, *P. M. H. S.*, Series 2. Vol. 43 (May 1910): 639-40.

was soon known through the city, and next morning the principal inhabitants of New York came to pay their respects and congratulations; among others, Governor Clinton, Dr. Prevoost, Bishop of New York, Mr. Osgood, late envoy to Great Britain, the heads of the college, most of the principal merchants, and deputations from the corporate body and other societies. No man in any public capacity,

wrote Wansey, "could be received with more respect than he was."[143] "It must afford sincere gratification to every well wisher to the rights of man," declared an editorial in the *American Daily Advertiser*,

that the United States of America, the land of freedom and independence, has become the asylum of the greatest characters of the present age, who have been persecuted in Europe merely because they have defended the rights of the enslaved nations.

The name of Joseph Priestley will be long remembered among all enlightened people. . . . His persecutions in England have presented to him the American Republic as a safe and honourable retreat in his declining years: and his arrival in this City calls upon us to testify our respect and esteem for a man whose whole life has been devoted to the sacred duty of diffusing knowledge and happiness among nations.[144]

It was a reception which Priestley, accustomed to a very different treatment at the hands of his fellow-countrymen, even if he professed to find it "rather troublesome," could not but enjoy. "It shews the difference of the two countries," he wrote. "With respect to myself, the difference is great indeed. In England, I was an object of the greatest aversion to every person connected with Government; whereas here, they are those who show me the most respect." The general difference between England and America was not, he said, to be expressed,

and whether it be the effect of general liberty, or some other cause, I find many more clever men, men capable of conversing with propriety and fluency on all subjects relating to government, than I have met with any where in England. I have seen many of the members of Congress on their return from it, and, without exception, they seem to be men of first-rate ability.[145]

"As to the government, it is nearly every thing we can wish," he wrote to Belsham, "and the few imperfections will be easily removed when it is the general interest and wish that they should be so; and here the majority bear rule."[146]

---

[143] Priestley, *Works*, I.2.234; D. J. Jeremy, ed., "Henry Wansey and his American Journal," 84, 85. Samuel Prevoost (1742–1815), the Episcopalian Bishop of New York, had been a pupil of Jebb's at Cambridge, and was sympathetic towards the tenets of Unitarianism. For Priestley's subsequent visit with Prevoost, cf. Wansey, "Journal," 128, and also below, n. 146. But cf. below, n. 147 for the universal failure of the churches in New York to allow Priestley to preach from their pulpits.

[144] *American Daily Advertiser*, 5 June 1794, cit. M. C. Park, "Joseph Priestley and the Problem of Pantisocracy," 3.

[145] Priestley to Lindsey, 6, 15 June 1794, *Works*, I.2.246, 255–9.

[146] Priestley to Belsham, 16 June 1794, ibid., I.2.259–61; cf. also Priestley to Wilkinson, 14 June 1794, W. P. L.; Mary Priestley to Belsham, 15 June 1794, ibid., I.2.236–7; and Wansey, "Journal," 127, for the dinner at the Osgoods, at which the Priestleys, Genet, and Prevoost, were present: "We had much interesting conversation after dinner, especially on political subjects."

"Almost every person of the least consequence in the place has been, or is coming, to call upon me," Priestley wrote to Lindsey. And he reported with great optimism also to Belsham of the prospects of employment in the colleges of America for his Unitarian friends. "The harvest truly is ready, and you must send us labourers." To Lindsey he wrote, however, of the prejudices still existing. "The preachers, though all civil to me, look upon me with dread." He had, he wrote, been asked by none of them to preach from their pulpits. But this, he firmly believed, would eventually prove to be to his advantage: "Several persons express a wish to hear me, and are ashamed of the illiberality of the preachers." He would, he wrote, "immediately" print his small pamphlets, "and wherever I can get an invitation to preach, I will go. With this view," he added,

I shall carefully avoid all the party-politics of the country, and have no other objects besides religion and philosophy. Philadelphia will be a more favourable situation than this, and there I shall make a beginning. It will be better, however, to wait a little time, and not shew much zeal at the first; and as my coming hither is much talked of, I shall reprint my Fast and Farewell Sermons.[147]

Priestley enclosed in this letter, however, for, as he expressed it, the amusement of his friends, "copies of some addresses and my answers, and also some letters from persons who are of a party opposite to the addressers, but equally friendly to me." His recognition that in America he was an inescapably political figure, whose actions were of no little significance, is implicit in his comment to Lindsey, that he found that he had "given as much satisfaction" to the members of the opposite party, "by the caution I have observed in my answers, as to the addressers, who, however, I believe, are now well satisfied that I do not openly join any of their societies, though at first I am informed they were very desirous of it." And he described for his friends, with his usual mastery and already considerable knowledge of the issues, the party alignments of America:

The parties are the Federalists and Anti-Federalists, the former meaning the friends of the present system, with a leaning to that of England, and friendship with England; the latter wishing for some improvements, leaning to the French system, and rather wishing for war. With a little more irritation, the latter will certainly prevail. They are now, I believe, by far the most numerous, especially in the country, though the others prevail in the towns, especially here.[148]

By the time Priestley wrote this letter, he had felt it necessary, in a letter to John Vaughan in Philadelphia, to explain his political position. He assured Vaughan "that the conduct that you wish me to pursue is the very same I had prescribed to myself before I left England." He asserted, as he so often had done in England, that he had been "thro life as little

---

[147] Priestley to Belsham, 16 June 1794, Priestley to Lindsey, 15 June 1794: "Time is necessary, and I am apt to be too precipitate. I want your cool judgement." Cf. also Wansey, "Journal," 1, 15 June 1794, 84, 128; and E. M. Wilbur, A History of Unitarianism in Transylvania, England, and America (Harvard Univ. Press, 1952), 395–6.
[148] Priestley to Lindsey, 15 June 1794.

as can well be supposed of a political character, having only been an advocate for general liberty, & a free representation of the people as the foundation of it"; and that he had "not so much as heard the names" of the political parties of America before his arrival in that country, and was "not disposed to make much enquiry about them." He was not, he stated, as he had in his Fast Day Sermon, averse to political associations as such. They were, he declared, in Godwinian terms, of value if they promoted political discussion and general curiosity. He believed, however, that in America the government was "so fundamentally good"–"without Bishops . . . without nobles & without a king," that he was sure that "whatever be imperfect, & requires amendment, will, I doubt not, in due time, find it." "As to myself," he added, "I have seen & felt so much of the greater abuses of government, that I shall perhaps be even too little attentive to smaller ones. For," he added, in implicit endorsement of some kind of opposition to the federal government, "these ought to be narrowly watched lest they should lead to greater." This letter, Priestley assured Vaughan, he would be happy, "lest any person should have been led to mistake my views and principles," to have made "as public as you please."[149]

Priestley did not mention in this letter the Addresses from the Societies in New York, two of which he had received and replied to by the time that he wrote it.[150] But it was almost certainly some such development that Vaughan had feared – and in this he was not alone. "A party is endeavouring to make a merit to themselves of your weight and influence," another writer from Philadelphia warned Priestley in the *Gazette of the United States*, urging him not to coalesce "with any party whatever." If he did so, he would only diminish his fame: if, however, Priestley remained in America a figure unconnected with any party warfare, "your private virtues, your industry in the pursuit of knowledge useful to mankind, will render your name respected as Franklin's."[151]

The Addresses from the Democratic Societies in New York to Priestley on his arrival in America were, as he himself clearly recognized, designed to attract his valuable and prestigious support to their efforts to rouse popular opinion against the prevailing inclination of the American government towards an alliance with England, and to encourage all those who wished well to the cause of France. The "multiplied oppressions" which characterised the English government, "the huge mass of intrigue, corruption and despotism" which were a feature of all the "governments of the old world," were deplored by the Democratic Society, as a base combination, "to prevent the establishment of liberty in France, and to effect the total destruction of the rights of man."[152] The Republican natives of

---

[149] Priestley to J. Vaughan, 8 June 1794, A. P. S., Misc. Mss. Colln., and Appendix. John Vaughan's letter to Priestley, which must have expressed a certain apprehension, has unfortunately not apparently survived.

[150] Cf. Priestley to Lindsey, 6 June 1794, *Works*, I.2.246.

[151] Priestley, *Works*, I.2.248–50; and cf. also Wansey, "Journal," 86.

[152] Priestley, *Works*, I.2.247–8; and P. S. Foner, *The Democratic-Republican Societies, 1790–1800*, 182–3.

Great Britain and Ireland similarly welcomed Priestley as a fellow opposer of "a corrupt and tyrannical government." They lamented the fatal apathy and despotic measures into which it had fallen, and gave thanks for the asylum provided by America "not only from the immediate tyranny of the British government, but also from those impending calamities which its increasing despotism, and multiplied iniquities, must infallibly bring down on a deluded and oppressed people."[153]

To these Addresses Priestley was to reply, in the inimitable words of Cobbett—who shortly launched his assault upon him for thus effectively demonstrating his willingness to enter the politics of America by answering them at all—with "sigh for sigh, and groan for groan."[154] For if Priestley, as he wrote to Lindsey, was careful not to join "openly" with the societies—and if, indeed, the Democratic Society was considerably piqued by his failure to unite with them—nevertheless in his replies he did enthusiastically and without reserve endorse their sentiments. "The wisdom and happiness of republican governments, and the evils arising from hereditary monarchical ones, cannot appear in a stronger light to you than they do to me," he declared. He lamented the degenerate and tyrannical state of Europe, and viewed "with the deepest concern" the prospect of "those troubles which are the natural offspring of their forms of government." He looked forward, as he said to the Democratic Society, to "that protection from violence which laws and government promise in all countries, but which I have not found in my own." He could not, he added, in a masterful piece of ambiguity, which Cobbett was to seize upon, "promise to be a better subject of this government than my whole conduct will evince that I have been to that of Great Britain." But he spoke of his unequivocal preference for the government of America to that of England, where, at present, "all liberty of speech and of the press, as far as politics is concerned, is at an end, and the spirit of intolerance in matters of religion is almost as high as in the times of the Stuarts." "I congratulate you, gentlemen," he wrote,

as you do me, on our arrival in a country in which men who wish well to their fellow-citizens, and use their best endeavours to render them the most important services, men who are an honour to human nature and to any country, are in no danger of being treated like the worst of felons, as is now the case in Great Britain.

He would be happy, he said, to welcome to America "every friend of liberty who is exposed to danger from the tyranny of the British government;" and he expressed his hope, "for the sake of the many excellent characters in our native country, its government may be reformed, and the judgements impending over it be prevented."[155]

---

153 Priestley, *Works*, I.2.251-3; and for other societies' Addresses, cf. Priestley, *Works*, I.2.241-2, 250-1; 254-5; Wansey, "Journal": Appendix, 150-5; and also Priestley, *Memoirs* II. Catalogue, x.

154 W. Cobbett, *Observations on the Emigration of Dr. Joseph Priestley, and on the Several Addresses delivered to him on his Arrival at New York* (Philadelphia, 1794), 3-4; 28.

155 Priestley, *Works*, I.2.248; 253-4; Wansey, "Journal," 86.

Priestley's replies to the Addresses from the New York Societies were, by the time that he was writing to his friends in England, published in the newspapers, "and circulated thro the continent." "This," he wrote to Wilkinson, "is rather troublesome to me but could not be avoided."[156] It was, however, to be the means whereby his first public gestures in America were seized upon by Cobbett, who was then teaching English to French emigrés in Philadelphia, and heard the Addresses, and Priestley's replies, read aloud to him by one of his pupils from the New York newspapers. The indignation which Cobbett felt was immediately translated into his first, and highly controversial, political pamphlet— *Observations on the Emigration of Dr. Joseph Priestley*. "His landing was nothing to me," Cobbett later wrote,

. . . but, the fulsome and consequential addresses sent him by the pretended patriots, and his canting replies, at once calculated to flatter the people here, and to degrade his country, and mine, was something to me. It was my business, and the business of every man, who thinks truth ought to be opposed to malice and hypocrisy.[157]

Priestley, wrote Cobbett, in this first of many such diatribes to be directed against Priestley in the next few years, was far from the politically innocuous figure which his "canting profession of Moderation, in direct contradiction to the conduct of his whole Life," would suggest. Nor, he further asserted, were his claims to compassion in any way justified. He was, Cobbett wrote, in language as slanderous and philistine as it was tasteless and provocative, an inflammatory and system-mongering revolutionary, whose "fellow-labourers" in the cause in England had, he implied, with good reason, been despatched to Botany Bay by the government, and whose own conduct amply justified all that the mob of Birmingham had perpetrated. Priestley himself, asserted Cobbett, would not have hesitated to have involved England in the horrors inseparable from the establishment of the revolutionary system of government which he so openly admired in France. His object now, as his replies to the Addresses in New York served to prove, was to fan the flames of prejudice against his country in America. Foiled of his object by the vigilance of government in England, "the Doctor, disappointed and chagrined, is come here to discharge his heart of the venom it has been long collecting against his country."[158]

While undoubtedly aware, as his correspondence from New York makes clear, of the potential for discord from his presence, but certainly as yet unaware of the extreme reaction it was to provoke from Cobbett,[159]

---

[156] Priestley to Wilkinson, 14 June 1794; and cf. Wansey, "Journal," 85.

[157] G. D. H. Cole, ed., *The Life and Adventures of Peter Porcupine* (London, 1927), 43; D. Green, *William Cobbett, The Noblest Agitator* (London, 1983), 124-5; G. D. H. Cole, *The Life of William Cobbett* (New York, 1924), 55-8; and cf. Cobbett, *Observations on the Emigration of Dr. Joseph Priestley*, 3.

[158] Cobbett, *Observations*, 4-30.

[159] Cf. *Philadelphia General Advertiser*, 1 August 1794, for the first announcement of what was certainly Cobbett's *Observations*: "A little pamphlet has been published in Philadelphia, containing a violent attack upon the character of Dr. Priestley. The extreme inhu-

Priestley on 18 June set off, as Wansey recorded, from New York for Philadelphia. On 19 June his party, which included his son Joseph and his
wife, stayed "a few hours" at Princeton, where they excited some interest;
and on the same day arrived in Philadelphia. "Doctor Priestley arrived
in this city from New York on Thursday last," announced Bache in the
General Advertiser on Saturday 21 June: "He remains a few days among
us but proceeds to the settlement his sons have been making in Northumberland County, in this State."[160] The *Advertiser* devoted a considerable amount of space in its columns to printing the New York Addresses
and Priestley's replies; and a report, taken from the *Morning Chronicle*,
of the Farewell Sermon which he had delivered to his congregation at
Hackney.[161]

"With respect to religion," however, Priestley reported to Lindsey,
things were "exactly in the same state" in Philadelphia "as in New York.
Nobody," he wrote, "asks me to preach, and I hear there is much jealousy
and dread of me." Others, however, had "on this account the greater
desire" to hear him; and he was in no doubt that "a respectable Unitarian
society" could be formed in Philadelphia: "and I stand so well with the
country in other respects, that I dare say I shall have a fair and candid
hearing."[162] In Philadelphia, as in New York, Priestley, as he wrote to
Wilkinson, was received "with the most flattering attention by all persons of note." He was "much pressed to take a house and reside" in
Philadelphia—a pressure which he was constantly to resist.[163] It was
in Philadelphia in 1794 that Priestley was first introduced, by Chief Justice McKean, to Washington.[164] And if he received "only one" Address,
as he wrote to Lindsey,[165] in this highly politicised city—that from the
American Philosophical Society, presided over by David Rittenhouse—it
was nevertheless one which possessed great significance for him, and
to which he composed a forthright reply. "I am confident," he wrote,

---

manity," the communication continued, "of abusing this respectable stranger on his arrival
in America, the compliments to the British constitution . . . and the invectives against the
French revolution and the French nation, render this publication utterly unfit for the
meridian of the United States." And cf. Cobbett's Introduction to the Third Edition (Philadelphia, 1795) for the adverse reaction, both in England and America, which his pamphlet
initially provoked. Cf. also Priestley to Lindsey, 24 August 1794, Priestley Colln., Dickinson
Coll. (and Graham, "A Hitherto Unpublished Letter of Joseph Priestley," *Enlightenment and
Dissent*, forthcoming): "The most virulent pamphlet that I have yet seen is just published
here against me. I will send you a copy. I shall not notice it, but I hear that some friend will."

[160] Wansey, "Journal," 129; "A Brief Description of Joseph Priestley in a Letter of David
English to Charles C. Green," *Presbyterian Historical Society Journal*, 38 (1960): 124–7; *Philadelphia General Advertiser*, 21 June 1794. Although cf. the date given in Rush's Commonplace Book (below, n. 168), 229.

[161] *Philadelphia General Advertiser*, 5, 7, 10, 13, 16, 18, 21, 23, 24 June 1794. And cf. J. Tagg,
Benjamin Franklin Bache and the Philadelphia Aurora (Univ. of Pennsylvania Press, 1991), 131
and n. 37, for the prominence given to Priestley in the *Advertiser* in 1791-3.

[162] Priestley to Lindsey, 24 June 1794, *Works*, I.2.263–6.

[163] Priestley to Wilkinson, 27 June 1794, W. P. L.

[164] Penn. Hist. Soc., McKean Papers, III.2, Washington to McKean, 9 July 1794: "The
President will be at home tomorrow at twelve o'Clock, at which time he will be happy to
see the Chief Justice & Dr. Priestly."

[165] Priestley to Lindsey, 24 June 1794, *Works*, I.2.266.

. . . that from what I have already seen of the spirit of the people of this country, that it will soon appear that republican governments, in which every obstruction is removed to the exertions of all kinds of talents, will be far more favourable to science and the arts than any monarchical government has ever been.[166]

During this stay in Philadelphia Priestley became well acquainted with Rittenhouse, who was president also of the Democratic Society in Philadelphia, "and passed much of his time" in his family.[167] And in conversation with another of the leading members of the Society, Benjamin Rush—whose company, Priestley was later to write, he found the most congenial of any in America—Priestley did not hesitate to speak "in high terms of republican principles," and in praise of France. "He said," recorded Rush, "that laws or opinions governed in France, and not men. This was proved by the same measures going on after the death or flight of so many of their leading characters."[168]

During this month-long sojourn in Philadelphia Priestley also made the first of what were to be a continuous and abortive series of efforts to realise the investments which Wilkinson had made for him in the French funds: "Everything will be abundantly easy to me," he wrote to his brother-in-law,

if I can secure the property you generously gave me in the French funds, and I have taken the best measures that I can for the purpose. I have drawn up a memorial on the subject addressed to the National Convention and the French minister in this city sent it this day to Paris, accompanied with one of my own in its favour, and he gives me the greatest encouragement with respect to it.[169]

As he was shortly to write to Benjamin Vaughan, Priestley's main circle of acquaintance was that of John Vaughan and his federalist friends. But he was almost certainly already drawn to those who represented radical, emigrant, and Francophile opinion, and to the many prominent French exiles quartered in Philadelphia—among whom was now Talleyrand—forced, like himself, to flee from England.[170] It is likely that he was intro-

---

166 *Minutes of the American Philosophical Society* (Philadelphia, 1884): 223: "Dr. Rush reported that a number of the officers and members of the Society waited on Dr. Priestley." For Priestley's reply, cf. ibid., 224; and cf. also Brooke Hindle, *David Rittenhouse* (Princeton, 1964), 348; and Priestley, *Works*, I.2.261-3. For Priestley's election to membership of the Society, in 1785, see above, n. 63.

167 W. Barton, *Memoirs of the Life of David Rittenhouse* (Philadelphia, 1813), 438.

168 G. W. Corner, ed., *The Autobiography of Benjamin Rush: Mem. Am. Phil. Soc.*, 25 (Princeton Univ. Press, 1948): Commonplace Book, 3 July 1794, 231; and cf. Priestley to Rush, 14 September 1794, *Scientific Correspondence*, 140: "I have not in this country met with any person whose mind seems to be so congenial to my own"; and Corner, ed., Common Place Book, 229, note.

169 Priestley to Wilkinson, 27 June 1794, W. P. L. Mss.; and cf. Chaloner, 29, 33, 38-9.

170 For Talleyrand's arrival in Philadelphia in April 1794, cf. Adams to Jefferson, 11 May 1794, above, n. 12. For the letters of introduction which Benjamin Vaughan wrote for him, praising his political character and principles, cf. Benjamin Vaughan to John Vaughan, 20, 27 February 1794, A. P. S., B. V. 462. For Talleyrand's sojourn in Philadelphia until the summer of 1796, cf. J. L. Earl, "Talleyrand in Philadelphia, 1794-1796," *P. M. H. B.*, 91 (1967): 282-98; R. L. de Beaufort, *Memoirs of the Prince de Talleyrand* (1891: repr. New York, 1973), 175-87; K. and A. M. Roberts, eds., *Moreau de St. Méry's American Journey*, 176-218.

J.PRIESTLEY. LLD.F.RS.

FIGURE 7.   Joseph Priestley, engraving by G. Murray, prefixed to a pamphlet published
20 April 1794, entitled *Character of Dr. Priestley, considered as a Philosopher, Politician
and Divine.* By permission of the President and Council of the Royal Society.

FIGURE 8. David Rittenhouse (1732–1796) by Charles Willson Peale. Courtesy of the American Philosophical Society.

FIGURE 9.   Benjamin Rush (1745–1813) by Thomas Sully. Courtesy of the American Philosophical Society.

FIGURE 10. View of Northumberland, early nineteenth century. Courtesy of the Historical Society of Pennsylvania.

duced to the Irish emigrant bookseller and political activist, Mathew
Carey.[171] And, as one of his letters to Lindsey makes clear, he did while
in Philadelphia certainly contact another emigrant publisher, the radical
bookseller Thompson, whom he had known in Birmingham. For publication by Thompson he wrote a Preface for an American edition of his
Appeal to the Public, copies of which, as well as reprintings of his Fast and
Farewell Sermons, he was shortly, as he had written to Lindsey, to distribute. "In them," he wrote, on sending some of these publications to
John Adams, "you will see my reasons for leaving England, and I hope
you will approve of them. You will see," he added with his usual disarming candour, "that I do not come hither from choice."[172]

It was while he was still in Philadelphia that Priestley heard, as he
described it to Lindsey, "the latest advices from England especially those
relating to Mr. Stone and the Corresponding Societies." They would, he
feared, "lead to much mischief." And, he added, with great frankness,
"I think however that if I had continued in England I could not have
escaped being involved with some of my friends, and therefore I think
myself happy, in being where I am, and wish more of my friends were
with me."[173] "The suspension of the Habeas Corpus, the commitment of
Mr. Stone (for whom I feel much interested) and the sending Mr. Tooke &c.
to the tower" were, he was shortly to write also to Benjamin Vaughan,
"considered here as desperate measures, and a prelude to some great convulsion, which I dread." Shortly after hearing of the fate of his friends
in England, Priestley left Philadelphia to undertake the arduous journey

---

[171] For Carey, cf. *D. A. B.*, and Edward C. Carter, "The Political Activities of Mathew
Carey," Bryn Mawr Ph.D. Thesis (1962). In July 1793, Joseph Priestley, Junior placed an
order with Carey: Penn. Hist. Soc., Lea and Febiger Colln., July 1793.

[172] Priestley to Lindsey, 15, 24 June 1794, *Works*, I.2.257, 266. (For Thompson, passage
only partially quoted in Rutt: cf. D. W. L. Mss., 24 June 1794: "Thompson is here, and super-
intends the office where it is printed. He will soon set up for himself." And Priestley to
Rush, 28 October 1794, *Scientific Correspondence*, 142.) And Priestley to Adams, 13 November
1794, Mass. Hist. Soc., Adams Papers, Reel 378. Cf. also Cobbett's later allegations that
Priestley "took good care to publish, and to distribute in great profusion, immediately upon
his arrival in Philadelphia," the "Preface to his farewell Hackney Sermon." Cobbett,
Porcupine's Works (London, 1801), IX.252. And Priestley to Lindsey, 24 August 1794, Priestley
Colln., Dickinson Coll., above, n. 159: "I shall introduce my small pamphlets as they are
printed at Philadelphia. The Appeal is among some of them."

[173] Priestley to Lindsey, 5 July 1794: D. W. L., passage omitted by Rutt. This passage, as
Priestley wrote it, is ungrammatical, and can cause confusion. Priestley wrote: "The latest
advices from England especially those relating to Mr. Stone, and the Corresponding Societies, as I fear they will lead to much mischief, I think however that if I had continued . . ."
C. Bonwick, in "Joseph Priestley, Emigrant and Jeffersonian," 18, uses the passage to come
to his conclusion that Priestley "feared the brutality and licentiousness of the lower classes
and disapproved of the efforts of the English Corresponding Societies to promote working
class interests during the 1790s." While the first part of this statement is certainly true (cf.
above, n. 118), the latter conclusion is, I think, open to question (cf. above, nn. 107, 149,
and also "Revolutionary Philosopher, Part II," 23–4). What is undeniable is that Priestley
recognised a situation of mounting political tension in England, realised that the arrests
by the government might lead to "much mischief," and was fully aware of the danger he
would have been in, as being effectively implicated in many of the reformers' activities, had
he remained. For the report in the *Philadelphia General Advertiser* of 3 July of the interrogation of Benjamin Vaughan and others before the Privy Council, cf. above, n. 121.

to Northumberland. Immediately upon his arrival there he seems to have decided, as he wrote to John Vaughan, "to make our principal residence in Philadelphia," and was hoping that Vaughan was "looking out for a proper house for me." By the end of July however he wrote to Benjamin Vaughan that he hoped to divide his time between Northumberland and Philadelphia in much the same way as he had between Birmingham and London. "I can reside a month or two in this city during the sitting of Congress," he had written to Wilkinson, "which will, in all respects, answer as good a purpose as living constantly there."[174]

Priestley was still at this time hoping, as he wrote to Lindsey, that "when a few of my friends are come we shall build a unitarian chapel, and probably have a *College*"; and he was clearly anticipating with some pleasure the prospect of the "large settlement" which his sons and friends were preparing "about fifty miles farther up in the country," to which he proposed to remove when it was sufficiently developed. The isolation of Northumberland, however – it was, he wrote to his sister, "seemingly almost out of the world" – he found already hard to bear. "In this interesting state of things I find it irksome to wait a whole week for news, but there is no remedy," he lamented to John Vaughan: "We know but little more than we did when we left you of European affairs." "I could now give a great deal for a complete set of the Morning Chronicle," he wrote to Lindsey, "or any tolerable English newspaper tho ever so old. I hope Mr. Belsham will send me the Cambridge Papers. They would amuse me much. We have only poor extracts in the Philadelphia papers. It is a long time since we have had any accounts of Mr. Stone, or our friends in the Tower."[175]

In his first two months in America Priestley, while protesting his disinclination for active politics, and clearly interested above all in the state of affairs which he had left in England and Europe, nevertheless in his outspoken replies to the Addresses of the Democratic Societies in New York, and in private conversation in Philadelphia, had made no secret of his political inclinations. His standing was such that his name was toasted by at least one Democratic Society, in Charleston.[176] He was still at this time held in high esteem by John Adams, but the regard in which he was also held among the leaders of Francophile opinion in America, and the impact which his arrival in America had made, can be seen in a letter which Jefferson, now in retirement at Monticello, wrote to Rittenhouse early in 1795: "If I had but Fortunatus's wishing-cap, to seat myself

---

[174] Priestley to Benjamin Vaughan, 30 July 1794, W. P. L., and Appendix; Priestley to J. Vaughan, 21 July 1794, Penn. Hist. Soc., Dreer Colln.; Priestley to Wilkinson, 27 June 1794, W. P. L.

[175] Priestley to J. Vaughan, 21 July 1794; Priestley to Lindsey, 24 August 1794, Priestley Colln., Dickinson Coll., and above, n. 159; Priestley to Mrs. Crouch, 29 July 1794, Royal Soc. Mss., 655.3; and cf. also Priestley to Belsham, 27 August 1794, *Works*, I.2.270-3.

[176] P. S. Foner, *The Democratic-Republican Societies*, 392: the Palmetto Society in Charleston toasted the success of the French; "The venerable Doctor Priestley, May his patriotic sentiments ensure to him that fraternal affection in this country, which despotism has denied him in his own"; and "Muir, Palmer and others, martyrs in the cause of freedom."

sometimes by your fireside, and to pay a visit to Dr. Priestly, I would be contented; his writings evince that he must be a fund of instruction in conversation, and his character an object of attachment and veneration."[177] In a letter which Jefferson was to write many years later, after he and Priestley had become well acquainted, he described his wish, on Priestley's emigration, that Thomas Cooper had chosen a different place in which to settle:

How sincerely have I regretted, that your friend, before he fixed his choice of a position, did not visit the vallies on each side of the blue ridge in Virginia, as Mr. Madison & myself so much wished. You would have found there equal soil, the finest climate & most healthy one on the earth, the homage of universal reverence & love, & the power of the country spread over you as a shield.[178]

In the letter which Priestley composed shortly after his arrival in Northumberland for Benjamin Vaughan, the gratifying reception which he had been accorded in America, the interest in the politics of his adopted country which clearly came so naturally to him, and his preference for those who were opposed to the policies of the administration were made very clear. "I have seen all the principal people," he wrote, "and also persons who may be said to be in the opposition. I take no part in the politics of the country, and consort chiefly with your brother, and his friends who are warm friends to *government*, as the phrase would be in England. I perceive," he added, however, "that the opposition is very considerable, and I am persuaded does not consist, as your brother will have it, of ill-intentioned men. They are called," he wrote,

Anti-federalists, and object principally to the *excise laws*, and *funding-system* founded on a *national debt*, which they wish to have discharged, while others avow a liking of it, as a means of creating a dependence on the governing powers, which they think is wanting in this country tho it has grown to a dangerous excess in England.

He discussed the inflationary effects of paper money, and some of the results to which this would inevitably lead: "It will put a stop to all emigration, except of labourers, and make manufacturing hazardous." He mentioned the trouble occasioned by the imposition of the excise, and his fears of "more mischief" arising from it; and he wrote also of the hopes of peace with England—"but the opposition lay less stress upon it." Only the success of French arms, it was believed, would procure a permanent peace. And the activities of the English in America, supplying arms to the Indians, were not an inducement to it. The poor laws, he regretted, were the same as in England: "Indeed, in many other things they seem to copy the English too closely, when they ought to take warning by the example."[179]

In two letters to John Vaughan also, Priestley revealed his close

---

[177] Jefferson to Rittenhouse, 24 February 1795, cit., Barton, *Memoirs of Rittenhouse*, 428–9.
[178] Jefferson to Priestley, 18 January 1800, Lib. Cong. Mss., Jefferson Papers, Ford, ed., *Works of Jefferson*, IX.96.
[179] Priestley to B. Vaughan, 30 July 1794, W. P. L., and Appendix.

(indeed alarmed) interest in the prevailing controversies in America, and his own instinctively democratic predilections. "By what I hear in this place the affairs in the western counties wear a more serious aspect than you in Philadelphia are aware of," he wrote, of the impending insurrection in Pennsylvania over the tax on whiskey.

I fear a *civil war*, and then I shall have got out of the frying pan into the fire. Would it not be better to give them up entirely rather than use compulsion, which may terminate as the war between Great Britain and this country, after doing much more mischief? Would not a tax on all lands cultivated or not be the fairest to satisfy all parties and prevent that monopoly by speculators that is so much complained of. Then upon lands reverting (to) the State the real settlers would have them on better terms. But,

he added, "I am no politician, I only wish the peace & welfare of the country I am come into."[180] "It is the wonder of all the friends of liberty in England," he had nevertheless written in another letter of some length,

that such a principle of taxation should be admitted here. Nothing makes Mr. Pitt's conduct more execrated than his extension of the excise. To make it productive, it must be oppressive, and it prevents all improvement in manufactures: for whatever any man does must be open to the exciseman. I have known several persons give up their business rather (than) be so watched, and have their improvements made public. Even a poll tax is, in my opinion, liable to fewer objections than an excise. The excisemen also in England favour the friends of government, and are rigorous with others, and so they will be here, for human nature is the same every where, and this must breed great discontent. In all this part of the country the excise is violently objected to. Nothing,

he concluded, in a sentence which demonstrates his fundamental political outlook very well,

can excuse an open opposition to the bringing of any tax laid by the representation of the people, but the authority of government is hazarded by forcing any thing that is very unpopular. People will chuse to do without government rather than pay so dear for it, and if they really chuse thus, they should be left to themselves.[181]

The confident tone of Priestley's first three months in America, his interest in her politics which, notwithstanding his refusal to participate in them, came so naturally to him, was early in September 1794 to be rudely interrupted by the sudden failure of the settlement on the Susquehanna, and the disappointment to many, as Priestley realised, in which this would result. The news was communicated in a letter to Lindsey of 14 September, and undoubtedly created consternation in England.[182] For Priestley himself, the news was a severe blow; and

---

[180] Priestley to J. Vaughan, 25 August 1794, *P. M. H. S.*, Series 2, Vol. 3 (June 1886): 16–17; and cf. also Priestley to Lindsey, 24 August 1794, Priestley Colln., Dickinson Coll., and above, n. 159.

[181] Priestley to J. Vaughan, 1 August 1794, A. P. S., Priestley Papers, B. P. 931.

[182] Priestley to Lindsey, 14 September 1794, *Works*, I.2.274; W. Vaughan to Wilkinson, 25 October 1794, W. P. L., and Appendix.

"what Mr. Cooper will do here I cannot imagine," he wrote. "We must all live on our means for some time yet to come."[183] He was at this very time offered the chair of chemistry at the "College in Philadelphia," and seriously considered accepting it. To the renewed consternation of his friends in England, however, he refused. "Had this proposal been made to me before the removal of my library and apparatus hither," he wrote, "the case would have been different; but this being now done, at a great risk and expence, I am, at all events, fixed for the remainder of my life."[184]

It was apparently after this damaging setback to his expectations, that Priestley, in November 1794, first made contact with his old friend and admirer John Adams, emphasising, as he had since his arrival in America, that "I only wish to be quiet, and pursue my studies without interruption, with the few advantages that I can expect in this country." He informed Adams of his great disappointment: "I came to this place with a view to a large settlement in which my son was concerned. But being come hither, and having, at a great risk and expence, brought my library and Apparatus hither, tho that scheme has failed, I cannot remove any more." He was relieved, he wrote, that the threatened disturbance in the west of the state had not materialised; and, in some apparent contradiction to his previous letter to John Vaughan, he wrote of his hope that "in consequence of the seasonable and vigorous exertions of government every thing will now be quiet." He himself, he wrote, had been in some danger from the extremes of political opinion prevailing among "the lower class of people . . . because I was understood to be against the erection of the liberty pole in this town, tho I have made it a rule to take no part whatever in the politics of a country in which I am a stranger, and in which," he repeated, "I only wish to live undisturbed as such."[185]

---

183 Priestley to Lindsey, 14 September 1794, D. W. L. Mss. for passage omitted in Rutt. Cf. also J. Priestley, Jr., to J. Watt, Jr., 20 November 1794, 10 November 1795, B. R. L.; and D. J. Jeremy, ed., "Henry Wansey's American Journal," 79–80, for the failure of the settlement.

184 Priestley to Lindsey, 14 September, 12 November 1794; Priestley to Belsham, 14 December 1794, *Works*, I.2.274, 280, 283; Priestley to Rush, 14 September, 28 October, 3, 11 November 1794, *Scientific Correspondence*, 139–45. And cf. also McKean to Priestley, 12 November 1794, Penn. Hist. Soc., McKean Papers, for the formal offer of the chair to Priestley; and also Lindsey to Freeman, 23 March 1795, Penn. Hist. Soc., Gratz Colln.: "I am concerned at his having declined the offer of the chemical professorship at Philadelphia, as it is so much better a sphere for him to act in than Northumberland, and furnishing opportunities of being more extensively useful." From a letter which he had recently received from young Priestley, however, Lindsey had heard that "the place of Professor was not yet filled up, the electors seeming still to wait for his father's acceptance of it, which," wrote Lindsey, "I ardently wish may be the case." In what was to be a crucial decision, however, Priestley refused all solicitations for the chair in Philadelphia, and effectively confined his sphere of operations to Northumberland. (For James Freeman [1759–1835] the first Unitarian Minister in America, cf. D. A. B.)

185 Priestley to Adams, 13 November 1794, Mass. Hist. Soc., Adams Papers, Reel 378; and cf. also Priestley to McKean, 16 November 1794, Penn. Hist. Soc., McKean Papers: "I had no other view in coming into this country than to find a peaceable and quiet retreat for the short remains of life, and hope I shall not be disappointed in my expectation. For some time, however, the appearances in this neighbourhood were very alarming. But by the seasonable and vigorous exertions of government we shall now I hope be quiet. I hope also that, in consequence of Genr. Wayne's victory, we shall not have much to fear either from the Indians or the British."

His protestations of a wish to live in peace and quiet notwithstanding, it was undoubtedly a sense of isolation which Priestley noticed most strikingly in the first few months of his life in Northumberland—"remote as I live from the busy world," he wrote to Lindsey, comparing his situation to that of a monkish retreat. He more eagerly than ever awaited news from England, and commented upon the extraordinary achievements of the French armies as they once more advanced into Flanders. The "Cambridge papers" which his friends sent to him were, he wrote, "a great feast for me. I shall hope to see more in due time. What great things are now depending in the course of Providence!" But, he wrote, he was "much affected by the state of things, and the danger that my friends are in. I hope, however, from your account," he wrote to Lindsey, "that Mr. Stone is not in much danger. Our last news is the taking of Bruges and Ghent, and, it is said, of Ostend, by the French." "We often wish to know what is doing in England," he wrote, "but we must wait with patience."[186] Early in November Priestley was further affected by the news of the capture of his close friend William Russell by the French, on the high seas, and his subsequent imprisonment, with his family, in Paris. "I was expecting to see him every day," he wrote to Wilkinson. And to Lindsey he described the measures which he had lost no time in taking on his friend's behalf:

As soon as I heard of it, I wrote to the French minister, with whom I am acquainted, particularly to shew the connexion of his case with my own, and the gratification that it would be to the English ministry that such a person as Mr. Russell should suffer as he has done.

The power of the French at sea was, he thought, "most astonishing." "A new state of things," he was sure, "is certainly about to take place, and some important prophecies, I believe, are about to be fulfilled." "It is, indeed, Sir," he wrote in a second letter to Adams,

as you observe *an awful crisis* in which we live. . . . But I flatter myself that since *truth* and *right* have a great advantage over their opposites, the present conflict of opinions, and of arms, will terminate in a better system than any that has hitherto prevailed. But the struggle I fear will be extensive and dreadful. Happy they who, looking to an overruling providence, can calmly wait the issue, endeavouring to lessen prejudice and violence on all sides, and contribute what they can to enlighten the minds, and improve the morals of their fellow creatures.[187]

---

[186] Priestley to Lindsey, 14 September 1794, D. W. L. Mss., passage partly omitted by Rutt, *Works*, I.2.274–5; to Lindsey, 16 October, 12 November 1794, ibid., I.2.278, 281; and cf. to Lindsey, 20 December 1794, ibid., I.2.285: "I have great pleasure reading the Cambridge Intelligencer, which you are so good as to send me. It is all the English newspaper that I see. If I can forward it without expense, I shall certainly send the packet to Mr. Toulmin." For Toulmin's arrival in Kentucky, cf. below, n. 204.

[187] Priestley to Wilkinson, 12 November 1794, W. P. L; Priestley to Lindsey, 12 November 1794, *Works*, I.2.279–80; Priestley to Adams, 29 November 1794, Mass. Hist. Soc., Adams Papers, Reel 378. For the capture of the Russells at sea by the French, and their imprisonment in Paris, cf. Jeyes, *The Russells*, 61ff; and also W. Russell to C. Vaughan, 15 November 1794, 31 January 1795, Charles Vaughan Papers, Bowdoin Coll.

FIGURE 11.   John Adams (1735–1826) by Gilbert Stuart, 1798 and 1815. Courtesy of the National Gallery of Art, Washington. Gift of Mrs. Robert Homans.

FIGURE 12.   Thomas Jefferson (1743–1826) by Thomas Sully. Courtesy of the American Philosophical Society.

Priestley's situation in the winter of 1794-5 was, as he remarked with candour to his many correspondents, very "distant from my original views." But, as he said to McKean, on refusing the offer of the chair of chemistry in Philadelphia, "I must sit down where I am, and make the best of my situation." If his adherence to the cause of France, and his perception of where, in theory at least, he stood in the politics of America, remained constant, and his appetite for political news as eager as before, nevertheless his many assertions at this time that he wished to have nothing directly to do with the politics of America must be believed. And one further development, arising out of the circumstances in which he had left England, must have served to strengthen his conviction that although, as he wrote, "my time here is far from passing so agreeably as it did in England," yet that he was "very thankful for such an asylum," and that his role, as he had written to Adams, was to "calmly wait the issue."[188] For in the memorandum which he now composed at John Vaughan's request, Priestley revealed how close he knew he had been to a situation of great personal danger in England. The fears which he had already expressed for his friends, and, had he remained, for himself, were, as this document demonstrates beyond doubt, well founded. "All that I can now recollect," wrote Priestley, describing the circumstances surrounding John Hurford Stone's communication with his brother in England—which had resulted in the arrest and imprisonment of the latter, and the precipitate flight of Benjamin Vaughan—

. . . is, that one sunday evening, when your family, and among them your brother Benjm, were taking a dish of tea with us at Hackney, Mr. Wm. Stone called and as he said he had news from France, I got his leave to introduce your brother. Mr. Stone then told us that a person, whose name I believe he did not mention, had brought him letters from his brother, but what particular news he brought I do not remember. He added that this person wished to know the state of parties in England; and whether there would be any probability of the French succeeding in an invasion, which was then much talked of. We agreed in desiring him to tell that person, that whatever might be the opinions, or wishes, of any party in other respects, no person would join the French in such a scheme, but that they might depend upon being opposed by the whole force of the kingdom.

In giving this advice, which we believed to be well founded, we thought we did service to our country, as it would tend to induce the leading men in France, if the opinion should come to their knowledge, to lay aside any scheme of an invasion, if they had intended it. Leaving the country presently after,

Priestley concluded, "I know nothing of what followed; but I am satisfied that neither Mr. Vaughan, or Mr. Stone, any more than myself, had any thought of encouraging the French in any hostile attempts upon England."[189]

---

188 Priestley to Lindsey, 22 February 1795, *Works*, I.2.295; Priestley to McKean, 16 November 1794; Priestley to G. Clark, 22 December 1794, *Works*, I.2.287.

189 Priestley to J. Vaughan, 8 December 1794; and cf. also same to same, 20 January 1795, A. P. S., Vaughan Papers, B. V. 462.

Throughout his years in America, Priestley was always to maintain, in his many letters to his friends, his fears for his country should the French invade. He frequently wished them, like himself, in a place of safety. That he feared the actual outbreak of revolution, whether in England, or, as on his arrival he apprehended in Pennsylvania, is clear. But that he increasingly believed it to be inevitable in his native country, and that he fully realised the dangers into which his opinions and actions there had led him—albeit at the hands of what he perceived to be an unjustly repressive government—is also clear. "All the accounts I receive convince me of the propriety of my leaving England," he wrote to Belsham in August 1794. And in this his friends in England concurred. "It is well for him," wrote Lindsey, "that he left this unworthy country most unworthy of him at the time he did; for such was the malice of his enemies, that they would have tried and perhaps succeeded to involve him in the accusations of sedition for which so many were taken up and tried." The same prejudices, Lindsey wrote, were still at work in England.[190] But in the ensuing years in America also, the prejudices against Priestley were to mount, and for the second time in his life he was to be faced with the unhappy prospect of an enforced removal from the country to which he was attached. For if he did not actively involve himself in the politics of America, nevertheless his political opinions, of which he made no secret—his open adherence to France in a period of great tension between that country and America, and his instinctively democratic predilections in the constitutional controversy resulting from it—were to place him once more in the arena of political debate.

---

[190] Priestley to Belsham, 27 August 1794, *Works*, I.2.273; Lindsey to Freeman, 23 March 1795, Penn. Hist. Soc., Gratz Colln.; and Priestley to Clark, 22 December 1794. Priestley's views both written and spoken were in many ways more circumspect than those of his son, who in November 1794 was writing to James Watt, Jr. of the "unpleasant situation" in which Watt stood, and that, "considering your state of mind with respect to politics," he really wished that he was in America. And yet, young Priestley added, were he in Watt's position, a bachelor, and unattached, "I should prefer staying in England & contributing my mite towards rooting those rascals Kings Priests & Nobles from the face of the earth. . . . I foresee with infinite pleasure the downfall of proud England the Establishment of the French republic, & the universal spread of equal rights & equal laws." J. Priestley, Jr. to J. Watt, Jr., 20 November 1794, B. R. L. For Joseph Priestley, Junior's association with Genet while awaiting his father's arrival in New York in May and June 1794, cf. Wansey, "Journal," 77, 84. He was, as Thomas Cooper wrote to Wilkinson, "a very good friend to the french." (T. Cooper to Wilkinson, 16 April 1794, W. P. L.)

PRIESTLEY IN NORTHUMBERLAND
1795–1797

In the early months of 1795, the survivors of the projected settlement on the Susquehanna were above all preoccupied with adapting to circumstances very different from their original expectations. Throughout that year Priestley, in his letters to his friends in England, attempted to justify his choice of settlement and continuance in such an "unactive" sphere—as Lindsey had apparently described it.[191] In Northumberland, he wrote, he had "the leisure for my pursuits that I could not have in a populous town," and "I endeavour to make the most of it."[192] He was at this time engaged upon his *Church History*, working at it, as he wrote, "between five and six every morning."[193] And he was also, in the midst of much inconvenience, reassembling his scientific laboratory. "This place is inconveniently situated for carrying on my experiments," he wrote to Samuel Parker, in a letter which described the great damage done to his apparatus "in its conveyance hither, owing chiefly to injudicious packing," and the difficulty of fetching such new instruments as had by now arrived for him in Philadelphia: "I am sending a slay, which is our best method of conveyance in winter, to fetch them, and other things that are waiting for me. We shall soon have a stagecoach and stage-waggon to this place, which will remove one of the greatest inconveniences we labour under."[194]

Northumberland, as Priestley had written shortly after his arrival there, conspicuously lacked communications with the outside world. "A certain and ready communication with England would add greatly to my satisfaction in this country," he wrote to Lindsey in August 1794, "and we have expectation of being soon better in this respect, as a stage is about to be set up between this place and Philadelphia and a post three times a week." In October, however, he could still lament that

---

[191] Priestley to Lindsey, 17 June 1795, *Works*, I.2.305. For Lindsey's concern, cf. above, n. 184.; and cf. also Sarah Vaughan to C. Vaughan, 15 August 1794, Charles Vaughan Papers, Bowdoin Coll.: "They will be influenced by their Sons but I hope they will not tempt them beyond the reach of Society if they do it will be very wrong they have so long been in the habit of that kind of intercourse with literary men that a life that is recluded (?sic) will not do for him, indeed it will not do for Mrs. P. either . . ."

[192] Priestley to Rev. Lindsey, 22 February 1795, Priestley to Rev. Samuel Palmer, n.d. 1795, ibid., I.2.296, 286–7.

[193] Priestley to Lindsey, 12 July 1795, *Works*, I.2.311; and cf. also to Belsham, 3 August 1795, ibid., I.2.313: "A year or more, applying to it as I now do, will enable me to complete the whole." And cf. Bakewell's account, cited by Rutt, ibid., I.2.311–12, note: "I always found him up and writing, when I went to the house, which seldom exceeded six o'clock in the morning."

[194] Priestley to Samuel Parker, 20 January 1795, Schofield, *Scientific Autobiography*, 284–5, 367; and also to Wedgwood, 17 March 1795, ibid., 285–6.

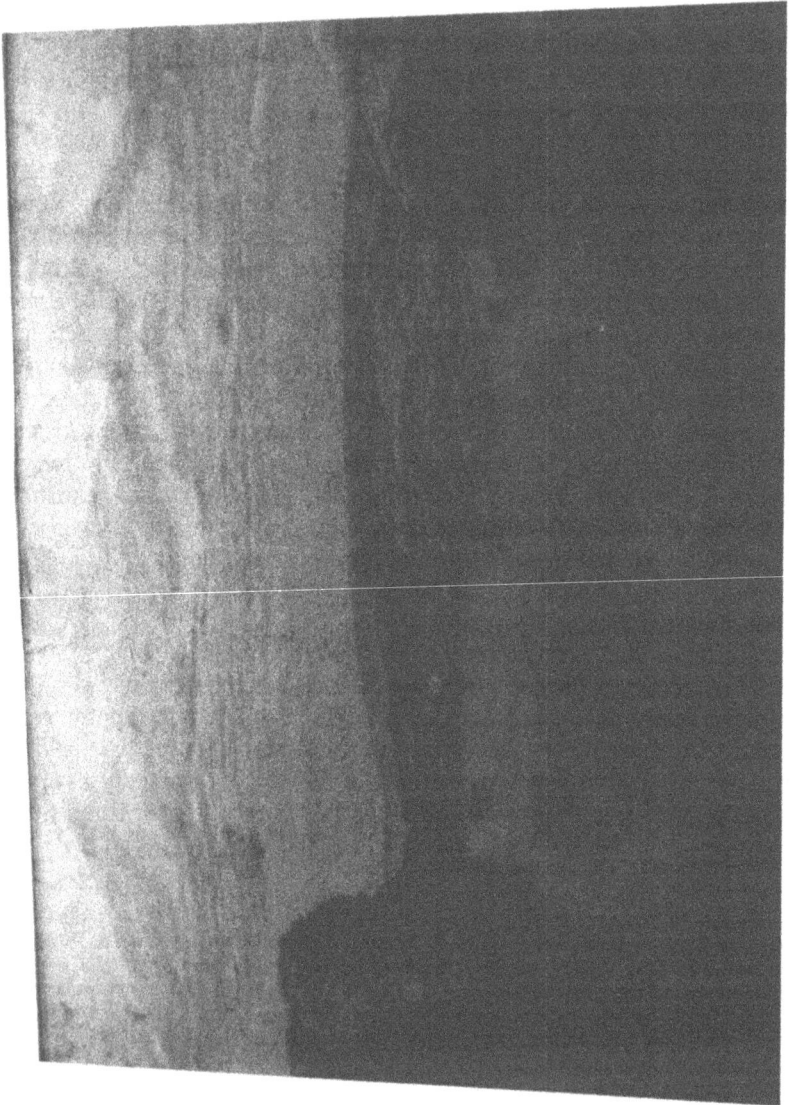

FIGURE 13. View of Northumberland, c. 1830, with Priestley's house and grave marked. Courtesy of Warrington Public Library.

FIGURE 14. "Plan of Dr. Priestley's House and Grounds at Northumberland in the State of Pennsylvania and their S. E. and N. W. Elevations. July 1800." By permission of the President and Council of the Royal Society.

By Prejudice and Error forced to roam,
Here found the exiled sage a distant home,
Here, mild Religion bade his troubles cease,
And active Genius earned the mead of peace.

BIRMINGHAM, RADCLYFFES & C⁰
1833.

FIGURE 15. Priestley's house in Northumberland, engraving, 1833, from the centenary
edition of his *Memoirs*. Courtesy of Special Collections, Dickinson College.

FIGURE 16. Portrait of Joseph Priestley in America, n.d., artist unknown. By permission of the President and Council of the Royal Society.

the greatest inconvenience attending this situation is a want of ready communication with Philadelphia. There are no stage-waggons; and the only method of sending heavy goods is by land in the waggons that carry corn to Middletown, on the Susquehannah, and thence by water hither; and the water is so low at this time of year, that it is not navigable. It is expected to rise a little towards the end of this month; but the best time for it is in the spring, and till midsummer; but then there are few waggons going to Middletown.[195]

In November he wrote to John Adams that "one of the greatest inconveniences I find here is the want of an easy communication with Philadelphia." And in reply to an offer from the latter "to assist me with respect to arrangements in the *Post Office*," he outlined the difficulties of the situation: "I am persuaded," he added,

that if the State would undertake to bring the letters to this place, if the post was regular, and the charge moderate, it would answer very well. At present the charge is so high, and the conveyance so uncertain, and tedious, that all persons take every opportunity of sending letters by private hands; when, if the case was different, they would all prefer the regular post. . . . Could we have a Coach . . . to carry *parcells*, and passengers, as well as letters, it would be a great convenience and benefit to the country, and in time would pay for any reasonable expence attending it. We sometimes talk of petitioning the legislature on the subject. Could you give us any assistance in the business, you would confer a great obligation on one who was so much interested in the conveyance of letters and small parcells.[196]

In January, Priestley wrote with some optimism to Lindsey. "In time, I am persuaded, this must be one of the finest situations upon this continent. But at present we labour under several considerable disadvantages, tho we have the prospect of the speedy removal of some of them. We shall very soon have a better communication with Philadelphia, and a diminution of our heavy postage, as well as a quicker conveyance."[197] He took the opportunity of his son's going to Philadelphia to write once more to Adams to request of him "particularly" his "kind assistance" in "the extension of the *post* from Reading to this place, with the reduction of the present high price of postage. We have a scheme of sending a *post coach* to Reading, and by the same means the *letters*, and *small parcells* may be brought. Several inconveniences arise," he added, "from the want of a better communication with Philadelphia, especially with respect to my philosophical pursuits." Later in that year, however, he was still writing despondently to Belsham that "here we have the post only once a week, and European news, which is all that interests us, two or three months after your have it."[198]

---

[195] Priestley to Lindsey, 16 October 1794, *Works*, I.2.275–6; and cf. also same to same, 24 August 1794, Priestley Colln., Dickinson Coll., and above, n. 159.

[196] Priestley to Adams, 13, 29 November 1794, Mass. Hist. Soc., Adams Papers, Reel 378.

[197] Priestley to Lindsey, 19 January 1795, *Works*, I.2.289, and D. W. L. Mss. for passage omitted in Rutt.

[198] Priestley to Adams, n.d. (January 1795), Mass. Hist. Soc., Adams Papers, Reel 379; Priestley to Belsham, 18 June, 3 August 1795, *Works*, I.2.307, 313.

Throughout 1795, however, Priestley was still hoping that Northumberland, as "a central and most agreeable place,"[199] would, if its communications with the outside world could be improved, attract Englishmen of his persuasion to settle in the vicinity. "I do not despair of seeing a college on the most liberal principles established in this place," he had written to Belsham in August 1794, "and sometimes I give a little more scope to my imagination, and fancy I may see *you* at the head of the institution." To Benjamin Rush in September, in discussing the proposed chair of chemistry in Philadelphia, he wrote that, before he left England, some of his friends "had a scheme of founding a College" wherever he settled, "on the idea that it would be in a part of the Country not provided with any." He asked for Rush's advice as how best to proceed with such a plan; and in November assured him that he was not "at all anxious" about the chair in Philadelphia, "hoping we shall succeed in establishing a College in this place." "Of this no doubt is now entertained," he wrote to Lindsey, "and the person whose property the greatest part of the town is, has consented to give the ground to build it on."[200] In March of the following year he could still write to Wedgwood with confidence of the proposed enterprise:

The scheme of a large settlement for English emigrants, projected by Mr. Cooper, you will before this time have heard is given up, and on the whole, though it mortified me at the time, I am now not sorry for it. Great difficulty would have attended the carrying it into execution, and many would have been dissatisfied. Where I now am there is room for a few, as many as I wish to draw near me, and in a short time, as the place has uncommon natural advantages, it must grow considerable. Already it is determined to establish a College here, and we expect to raise the necessary buildings this very year. A handsome subscription has already been raised, and a petition is now before the Assembly of the State for a grant of lands for its support, and there is no doubt of its being duly attended to.[201]

In May Priestley wrote once more to Rush, thanking him "for the pains you have taken about our *Academy*." To Lindsey in the same month he wrote of his hopes that he could devote "whatever emolument I derive from the College we are about to establish in this place" to the implementation of his plan of founding a Unitarian congregation in Philadelphia. "I have written to my friends in Philadelphia," he reported, "to acquaint them with my resolution, saying I would appear among them, if at all, in my proper character of *a Christian minister*, and would not again be reduced to a state of disgraceful silence by the bigotry and jealousy of their preachers." "I could have wished to do for it what I did for

---

[199] Priestley to Lindsey, 12 July 1795, ibid., I.2.312.
[200] Priestley to Belsham, 27 August 1794, *Works*, I.2.271-2; and same to same, 16 June 1794, ibid., I.2.260-1; Priestley to Lindsey, 16 October, 12 November 1794, ibid., I.2.275-81. Priestley to Rush, 14 September, 28 October, 3, 11 November 1794, Schofield, 281-3; *Scientific Correspondence*, 139-45.
[201] Priestley to Wedgwood, 17 March 1795, Schofield, *Scientific Autobiography*, 285-6.

the College at Hackney," he added further of the proposed College in
Northumberland, "but I have so many demands upon me here that I
cannot well do it."[202]

In October Priestley wrote to his old Lunar companion, Withering.
"More than ever do I now regret the loss of the *lunar society*, where I spent
so many happy hours, and for which I found no substitute even in
London. Here I am quite insulated, and I promise myself, when my
house and laboratory shall be erected, to devote as much time to philosophical pursuits as ever I have done. Hitherto it has not been in my
power to do much, as I have only one room in my son's house for my
library and apparatus too." He had, however, completed a series of experiments on air, "and shall soon draw up a *sequel* to my pamphlet on that
subject for the society at Philadelphia." He was "chiefly employed in the
continuation of my *History of the Christian Church*, tasking myself, as I presume you do, so much every day. . . . Soon, however, I expect to be
employed in the instruction of youth, as a *College* is to be established in
this place, and I am appointed the Principal. The next spring we begin
to build, but at first our funds will be small. I wish we had a proper
person for teaching *Natural History*, including Botany. Almost everything
else I can, *pro tempore*, in some measure, teach myself."[203]

The prospect of the college in Northumberland, his hopes of spending
"two months annually in Philadelphia," and his optimism that he could
gather "as many as I wish to draw near me," was a valuable corrective
to the great isolation which Priestley, above all in his correspondence
with Lindsey, confessed that he felt in his first year in America. The letters which Lindsey and Belsham wrote were, he assured his friend, "the
greatest satisfaction that I have in this place." He wrote, however, of his
hopes that young Toulmin, who had settled in Kentucky, might, "if we
get a college here," become one of its tutors, and even that he might "see
several of my much-valued friends, though not yourself, on this side the
water."[204] By the late summer of 1795, however, Priestley feared that,
although Toulmin was far from satisfied with his lot in Kentucky, things
in Northumberland were not in such a state of forwardness "as to afford
ground to expect such society as that of his, or any other friends from
England."[205] He was by now lamenting also that "the present governing
powers have shewn a ridiculous jealousy of democratical emigrants, and,

[202] Priestley to Rush, 22 May 1795, *Scientific Correspondence*, 145–6; Priestley to Lindsey,
17 May 1795, D. W. L. Mss., passage omitted in Rutt, *Works*, I.2.302-3.

[203] Priestley to Withering, 27 October 1795, *Scientific Autobiography*, 287–8; *Scientific
Correspondence*, 148–51; A. P. S., Priestley Papers, B. P. 931.

[204] Priestley to Lindsey, 10 February 1795, *Works*, I.2.293-5. For Toulmin's arrival in Kentucky, cf. Lindsey to Freeman, 23 March 1795, Penn. Hist. Soc., Gratz Colln. For Priestley's
continuing hopes of persuading him to come to Pennsylvania, cf. his letter to Lindsey, 12
July 1795, *Works*, 1.2.312, and T. Cooper to Watt, Jr., 5 July 1795, B. R. L.: "Toulmin does
not like Kentucky. Dr. P. recv'd a Letter from him a day or two ago, wishing to settle
hereabout."

[205] Priestley to Lindsey, 12 August 1795, *Works*, I.2.314.

from a dread of them, as Mr. Adams acknowledges to me, they have, in the last congress, made naturalisation more difficult than before." He welcomed, he wrote, Lindsey's recommendation of the son of Professor Millar: "Had our college been established, I should have thought him a valuable acquisition. However, several of our zealous friends are of the aristocratical or governmental party in this country; and to them Mr. Millar's having left England for his attachment to the principles of liberty will be no recommendation." He himself had assured Adams that "I had no intention to be naturalized at all, but to live as a peaceable stranger. I can perceive however," he wrote, "that the democratical party is growing stronger, and will, in time, get the upper hand. Party spirit is pretty high in this country, but the constitution is such," he confidently asserted, "that it cannot do any harm."[206]

If it was impossible for Priestley to be wholly isolated from the prevailing social and economic conditions of America, and the political developments that were so much affecting his and his friends' prospects, yet it was, as he frequently wrote, only the news from Europe, and of his fellow countrymen, that held any great interest for him. In May indeed his despondency was such that he wrote to Lindsey: "Were it not for the concern I have for my friends, and the attention I give to the fulfilment of prophecy, I should take but little interest in the politics of Europe. Here we are, as it were, out of the world, and begin to give but little attention to it." In the following month, however, on the receipt of a package of pamphlets and newspapers from England, his waning interest revived:

I can hardly give you an idea of the interest I take in every thing that comes from England, and how little in anything here. This is in a great measure, no doubt, owing to there being nothing very interesting now going forward here, every thing being quiet, and only in a silent, regular progression to a better state; whereas with you the greatest events may be expected, and things cannot continue as they are; and with the fate of England is connected that of Europe, and of the world.

The *Morning Chronicle* was, he wrote, "particularly welcome" to him:

I plainly perceive by it that the spirit of the people is getting up, and that things are approaching to the state they were in towards the close of the American war. I wish the issue may not be more calamitous. I am far from rejoicing in the distresses of my native country, and even those of my enemies in it; though I as earnestly as ever wish well to the cause of liberty, and, consequently, the success

---

[206] Priestley to Lindsey, 12 July, n.d. (1795), *Works* I.2.303–4, 310–12; and also Priestley to Rush, 22 May 1795; and Priestley to Adams, n.d. (January 1795), Mass. Hist. Soc., Adams Papers, Reel 379: "I have no wish to have any privilege of voting &c here, I have no thoughts of being naturalized, but propose to live as a peaceable stranger." For the new regulations for naturalisation passed in 1795, "stipulating that foreigners must reside in the United States for five years before they were entitled to citizenship," see J. M. Smith, *Freedom's Fetters: the Alien and Sedition Laws and American Civil Liberties* (Cornell Univ. Press, 1956), 23. For John Miller, who did emigrate to America in 1795, but whose career was cut tragically short by his premature death from sunstroke, see *D. N. B.*, in the entry for his father; and W. Barton, *Memoirs of Rittenhouse*, 423–4.

of the French. I rejoice in the change of measures that has taken place in that country, and wish it may be permanent; but, by our last accounts, another revolution was apprehended. By this time you know much more than we do here, and much more you will know before you receive this.[207]

America, Priestley wrote many times to his English correspondents, enjoyed great advantages from its "happy constitution of government, and a state of *peace* in consequence of which the country enjoys an unexampled prosperity, the advancement in population, and improvements of all kinds, being beyond any thing that the world ever saw before." But, he wrote, although he had found "a happy *asylum* here," he could "consider it in no other light. I feel myself in a state of *exile*, and my best wishes are for my native country and my friends there."[208] There was, however, one concern which for him overrode all others in these early years in Northumberland, and which undoubtedly affected his thinking. "Though this country, and the government of it, are really almost every thing that I can wish from them, I shall always feel as a stranger," he wrote to Belsham, "but my sons will be at home soon."[209] His three sons, as he had informed Samuel Parker in January 1795, were settling themselves "in farms about me."[210] In December he had informed Lindsey, who had asked after his sons, that "Joseph has taken a house in this town, but has not yet got any farm of consequence. We have bought 300 acres of the cheaper kind of land," he added, "but Harry has taken a fancy to it, and so he is to have it, and to enter upon it next spring. It is yet to clear; but it is full of good timber, which will almost pay for the clearing." For William, who was still in Boston, he would endeavour to do similarly. "I am happy," he wrote, "that none of my sons seem to have any wish to be more than plain farmers and what their own industry will make them. If they persevere they cannot fail to do well, and live comfortably, tho they cannot attain to affluence. They have," he characteristically added, "all a turn for reading, which will render them much superior to the generality of American farmers, and indeed the generality of professional men in this country."[211] To his friends in England Priestley described his own work in the fields with his sons — how "even I sometimes take my axe or my mattock, and work, as long as I can, along with them."[212] "On the whole," he wrote,

---

[207] Priestley to Lindsey, 17 May, 17 June 1795, *Works*, I.2.302–3, 305–6.

[208] Priestley to Samuel Parker, 20 January 1795, Schofield, *Scientific Autobiography*, 285.

[209] Priestley to Belsham, 3 August 1795, *Works*, I.2.313.

[210] Priestley to Samuel Parker, 20 January 1795, Schofield, *Scientific Correspondence*, 285; and cf. also Priestley to Adams, n.d. (January 1795): "My sons are settling as farmers in this neighbourhood; so that I have a prospect of being as happily situated here as I can expect to be anywhere."

[211] Priestley to Lindsey, 20 December 1794: D. W. L. Mss., passage partly omitted in Rutt, *Works*, I.2.283–5. And cf. also same to same, 17 May 1795, D. W. L. Mss., passage omitted in Rutt, on Priestley's considerable outlay of expenditure in settling his sons on the land.

[212] Priestley to Lindsey, 5 April 1795, *Works*, I.2.300–2; and also same to same, 12 July 1795: "Two or three hours I always work in the fields, along with my son. The weather beginning to be hot, I do this early and late."(ibid., I.2.311).

I have much reason to be thankful with respect to them all. Their minds are perfectly conformed to their situation. They are very frugal, and industrious, not at all ashamed to anything that their employment requires. Harry drives his own horses and cart, and works with his men like one of them, and here there is little difference between master and servant. Indeed, these terms are unknown.[213]

In August, however Priestley revealed how greatly he felt the hardships inseparable from their new way of life: "I am more and more satisfied that my coming here was necessary for the settlement of my *sons*. If William and Harry had had no (house) besides their own log houses, they would have despaired."[214]

In one further respect Priestley was much affected by the conditions of life in America, and inevitably drawn to comment upon its implications. "Nothing is yet done towards building my house," he lamented to Lindsey in April 1795. "It is next to impossible to get workmen, and the price of every thing is advanced one-third since we have come hither. This indeed is an alarming circumstance, and how far the evil will go we cannot tell. It is with most probability ascribed to the increase of paper money, which has only very lately made its way hither."[215]

If the Priestleys were adapting to their new mode of living with some difficulty, Thomas Cooper, whose relationship with them at this time was far from close, was, as he described it, "sufficiently engaged in procuring bread & cheese for myself. I have bought about 500 Acres of Land," he informed young Watt, "3½ miles from Northumberland. . . . The land stoney but well timbered: a brook runs through it, & the situation (answers) very well." He had, with Joseph Priestley, Junior, indulged in some land speculation, and had it not been for this,

I should have been cursedly ill off.—Since I was here before, the price of Land, provisions and Labour have nearly doubled. A rise so sudden and so permanent has astonished all the People here, who know not how to account for it: but the increased home population, the emigrations, the foreign demand for produce, & the doubling of the home Capital by the establishment of Banks, & the general increase of property, is quite sufficient to explain it.

He lived, Cooper continued, in "a Log House of one room below, & a room (under the roof) over it. To this I have added a room of Boards nailed to some Posts set in the ground where I live." Mrs. Cooper, he explained, "will not come into the Woods till I get her a House built which I will do when I get more Cash. . . . So many unthinking and unreasonable people have come out," he added, "that I have nearly enough of English Society; and I rejoice that those who have been disappointed have not settled here. The *generality* of the Land in this Neighbourhood is not rich," he admitted, "otherwise it would be a Paradise, but it

---

[213] Priestley to Lindsey, 12 July 1795, D. W. L. Mss., passage partly omitted by Rutt.
[214] Priestley to Lindsey, 12 August 1795, D. W. L. Mss., passage omitted by Rutt, *Works*, I.2.314–15.
[215] Priestley to Lindsey, 5 April 1795, ibid., I.2.300.

requires some skill & some Labour, to make the most of it. However," he bravely asserted, "I am satisfied with my Choice." But, he added, in discussing the troubled politics of England, "if it were not for my Family I should like to be with you, but all things considered I think myself better here, as I think you would be so too."[216]

At this waning stage of the fortunes of the settlement on the Susquehanna, when Priestley also was writing bravely to Lindsey that "I never before found such aid from a sense of religion as in my present circumstances, and a persuasion that both in coming to this country, and settling here, I acted from the best views, as things appeared at the time,"[217] the Duc de la Rochefoucauld Liancourt visited Northumberland, and recorded his far from entirely favourable impressions of the immigrants. Priestley, he wrote, maintained the same mode of living, and of dress— "the wig excepted"—as he had in England. "He frequently laughs at the world," wrote Liancourt,

but in a manner which clearly appears not to be from his heart. He spoke with great moderation of the political affairs of Europe, and in very mild expressions of England. He is now busied in the institution of a college, for which six thousand dollars have already been subscribed, and seven thousand acres have been assigned him, as a free gift. In this establishment, of which he has drawn up a prospectus, there is a president's place, doubtless intended for himself. Joseph Priestley, the eldest son, seems,

wrote Liancourt—apparently drawing the contrast with Priestley—"at present to be more engaged in industrious pursuits, than in political discussions." Thomas Cooper had

purchased some hundred acres of land, which he is at present clearing of wood, and preparing for cultivation. He is undoubtedly a man of parts, of a restless mind, ill adapted to find happiness in a retired rural life. In the account he wrote of America, it was certainly his design, to persuade colonists to join Dr. Priestley. In his manners, he affects at present a strong predeliction for American customs; and says, that he prefers his present mode of living to any other.

"He is suspected here," concluded Liancourt, "of aiming at a seat in Congress. In point of abilities at least, he would hold no mean rank among its members." Liancourt was not impressed, however, by the "cold and gloomy tempers" of the Priestley family. He recounted, with a good deal of inaccuracy, Priestley's refusal of the chair of Chemistry in Philadelphia; and he asserted that the hauteur and austerity of the Priestley

---

[216] T. Cooper to J. Watt, Jr., 4 July 1795, B. R. L. For Cooper's arrival at Northumberland late in November 1794, cf. J. Priestley, Jr. to Watt, Jr., 20 November 1794, B. R. L.; Priestley to J. Vaughan, 17 December 1794, A. P. S., Priestley Papers, B. P. 931. Cooper's years in Northumberland, he said in old age, were "a complete blank in his life." (E. A. and G. L. Duyckinck, *Cyclopaedia of American Literature* [1855], II.333). And Malone, his biographer, could only surmise, in the apparent absence of any private correspondence, on his activities during these years (Malone, *Cooper*, 81ff.). Cooper's letters to young Watt provide, however, a valuable and revealing chronicle of his first years in America.

[217] Priestley to Lindsey, 12 August 1795.

household had materially contributed to deter other English emigrants from joining them.[218]

Above all in his increasingly beleaguered situation in Northumberland – a philosopher cut off, as he himself recognised, from many of his sources of strength – Priestley longed for the company of those of his English friends who had, with him, fled from England as political refugees. In December 1794 he was hoping to see Benjamin Vaughan "in the spring"; and he was mortified on hearing, from both Samuel and William Vaughan, that although Benjamin was apparently then planning to embark for America, "that he will probably settle at Boston. I can only say that I wish he was nearer to me."[219] He was still expecting the imminent arrival of William Russell and his family, and indeed a remonstrance from his son to William Russell that it was felt by the party on the Susquehanna that the Russells – now released from captivity, and enjoying the social and political life of Paris – were not displaying a sufficient readiness to embark for America,[220] undoubtedly reflected the feeling of isolation which was very generally affecting the settlement in Northumberland.[221]

In September 1795 however, William Russell and his family finally arrived in America and, Priestley wrote, "my prospect is much improved." The Russells, he believed – as did Thomas Cooper – liked "the place so well as to settle in it." They had, wrote Priestley,

taken a house in Philadelphia, where he will reside in the winter, and where it is settled that I am to be his guest, as long as I choose to continue there, which will add much to my satisfaction. He will also have packages coming to him hither at least twice a year, and in these any thing may be sent to me with great convenience. Besides this, we shall soon have four new circumstances in our favour, a market, a stage-coach to Philadelphia, a corporation, and a college. The third, judging from your ideas, you may think to be no advantage; but here it is found to be a great one, as the majority can then bind the minority, in removing nuisances, and making regulations for the convenience of the whole. In these

---

[218] La Rochefoucauld Liancourt, *Travels Through the United States of North America*, 73–6.

[219] Priestley to J. Vaughan, 16 December 1794, 6 May 1795, A. P. S., B. P. 931. Although Priestley wrote in December 1794 that he had heard from Benjamin Vaughan, the first extant letter he wrote to him after 30 July 1794 (above, n. 179) would appear to be after his arrival in America: see below, nn. 272, 273.

[220] J. Priestley, Jr. to William Russell, 10 May 1795, Russell Papers, Penn. Hist. Soc., Bundle XXXI: "Northumberland is now the place of my father's residence as likewise of Mr. Cooper, my brother Harry, Mr. Forster & Mr. Darch. It is impossible for me to judge of your situation at Paris & I have no doubt, but that you have good reasons for staying there, but it has long been the wish of myself, & all your friends, that you should come to this country *as soon as possible.*" For the sojourn of the Russells in Paris, cf. S. H. Jeyes, *The Russells*, 116–55. And for Russell's purchases of land in France, cf. T. Cooper to (?T. and R. Walker), 29 September 1795, B. R. L.: "He has bought a Convent and an annexed estate in France, where he wanted much to settle in preference to coming to America, but tho' he is in love with the Country, & thinks well of the People, & is satisfied of the stability of the Government his Son & Daughters dislike it so much that they persuaded him to quit it for good."

[221] Priestley to Lindsey, 17 May, 17 June 1795, *Works*, I.2.302–3, 305.

respects, Mr. Russell will be a treasure to us; and his activity and public spirit is increased, rather than abated, since I saw him last.[222]

In these hopes, however, Priestley was to be sadly disappointed. The Russells, and in particular William Russell's son, Thomas, were horrified by the conditions in which Priestley was living. Priestley's house Thomas Russell thought "a mere hut in comparison with the one they lived in formerly." The climate he thought "unhealthy"; the soil infertile. Thomas Cooper's farm was "as barren, rude a farm as he could have selected." Had they been banished by law to such a situation, thought Thomas Russell, "they would have been right to make the best of their situation." But to retire there into voluntary exile, appeared to him extraordinary and unaccountable. In October the Russells departed for the preferred climes of New England. "This is a circumstance not a little unfavourable to me," Priestley wrote, "as the want of such society as I was used to in England is one of the greatest that I feel here."[223] And in that month also, he had the further disappointment to bear of hearing that Benjamin Vaughan had finally decided to settle on the family estates in Maine. "His being *here*," wrote Priestley, "would be a great addition to my satisfaction in the place. What I chiefly want is such society as his." He did, however, with his customary resignation, shortly accept this blow also. "I should much rejoice to see Mr. B. Vaughan," he wrote to Russell, "but it would be very unreasonable to expect it. He would enter into all my views, theological, philosophical, or political; but he must have a larger sphere than this place can furnish, and I am getting too old for him."[224]

It was in the autumn of 1795, when they were clearly demoralised by

---

[222] Priestley to Lindsey, 14 September 1795, *Works*, I.2.317; T. Cooper to (?T. and R. Walker), 29 September 1795, B. R. L. For Priestley's hopes for the incorporation of Northumberland, cf. his letter to Adams, n.d. (January 1795), Mass. Hist. Soc., Adams Papers, Reel 379.

[223] S. H. Jeyes, *The Russells of Birmingham*, 169, 197–9; Priestley to Lindsey, 9 November 1795, *Works*, I.2.320; and also Priestley to Russell, 10 November 1795, *Works*, I.2.321: "In a variety of respects, as well as liberality, we must yield to New England; but a train of circumstances have brought me hither, and it is impossible for me to remove, and our difficulties are not so great but we may get through them, though not so soon as if we had had the aid of your activity, as well as experience and public spirit. I shall certainly regret the want of your society, but I shall hope to meet you sometimes at Philadelphia, and perhaps you may again pay us a short visit here." Cf. also Priestley to J. Vaughan, 17 October 1795, H. C. Bolton, ed., *Scientific Correspondence*, 147; and Priestley to Lindsey, 6 December 1795, *Works*, I.2.325, on the Russells' decision to settle in Middletown. Part of Priestley's concluding sentence—a characteristically acerbic comment—was omitted by Rutt: "I shall, however, meet him in Philadelphia," Priestley wrote, "and by that time he may think as little of Middletown, as he now does of Boston. He is too fickle." Cf. however the letter from George Thatcher to James Freeman (below, n. 237) in which he describes Russell's attempts in the spring of 1796 when Priestley was staying with him in Philadelphia, to persuade his friend to visit Boston.

[224] Priestley to J. Vaughan, 28 October 1795, A. P. S., Priestley Papers, B. P. 931; Priestley to Russell, 10 November 1795, *Works*, I.2.322. For Benjamin Vaughan's eventual arrival in America, cf. below, n. 272. His wife had preceded him in 1795 with their children: cf. S. Vaughan to S. M. Vaughan, 17, 22 July 1795; S. Vaughan to B. Vaughan, 13 August 1795, B. V. 46 p; and B. Vaughan to J. Vaughan, August 1795, B. V. 462.1; Murray, *Benjamin Vaughan*, 349–50; Marvin, *Benjamin Vaughan*, 44–7.

their failure to attract Englishmen of substance to Northumberland, that the coolness between the Priestleys and Thomas Cooper, which was to affect their relations for some considerable time, seems to have developed. Shortly before his embarkation for America, Priestley had written to Cooper of his concern at hearing of the latter's growing religious unbelief:

and that you have moreover accustomed yourself to profane swearing in conversation; as I had flattered myself that in you I should have had an able friend, and ally, in the defence of christianity, and of religion in general. I am concerned on your own account, and that of the world at large; but I confess that I am also particularly concerned on account of our proposed connection in a settlement in America (which I hoped would be an asylum for my christian and unitarian friends), and for your intercourse with my sons, lest they should take the same turn.

Priestley was not, he wrote, attempting to deny to Cooper "such freedom in speculation as I indulge in myself"; nor did he wish to change his opinions: "But I deeply regret the event, and it will make me cross the Atlantic with a heavy heart."[225] This letter, Priestley wrote to Lindsey, Cooper "took very ill, and he seemed then disposed to enter into a public discussion of the subject with me. I shall (?not) decline it. I had rather it had been with another person."[226]

The growing prevalence of irreligion was for Priestley a cause of the greatest concern. On the Atlantic crossing—"though it was particularly inconvenient to write long hand"—he had "composed about as much as will make two sermons on the causes of infidelity." And on his arrival in Northumberland he wrote to Lindsey that he had "just seen *Paine's Age of Reason*, and shall probably make some remarks on it, in another letter to the philosophers in France. It is arrogant and absurd in the extreme."[227] In November he was writing to Lindsey of "some important additions to my 'Observations on the Causes of Infidelity,'" and also his "answer to Mr. Paine, which I find is well received in this country."[228] In February he could write that in enlarging his "Observations" he was "assisted . . . indirectly, by the conversations I sometimes have with Mr. Cooper on the subject." And that he was far from retracting entirely the "real esteem and affection," which in April he had declared that he felt for Cooper— "notwithstanding our great difference of opinion on subjects of the

---

[225] Priestley to T. Cooper, 6 April 1794 (and cf. above, n. 111 for the provenance of this letter).

[226] Priestley to Lindsey, 14 September 1794, D. W. L. Mss., passage omitted by Rutt, *Works*, I.2.273–5.

[227] Priestley to Lindsey, 6 June 1794, *Works*, I.2.245; and to Lindsey, 24 August 1794, Priestley Colln., Dickinson Coll., and above, n. 159.

[228] Priestley to Lindsey, 12 November 1794, *Works*, I.2.280. For Priestley's reply to Paine, which he published in the spring of 1795, cf. his letter to Belsham, 22 March 1795, *Works*, I.2.297; and also Priestley to Lindsey, 14 September 1795, *Works*, I.2.318: "I thought, till Mr. Russell informed me to the contrary, that you had got the edition of my answer to Mr. Paine, which was printed at Philadelphia. Mr. Russell says he left his copy with Mr. Paine at Paris, who promised to return it, but did not. He could not tell what he thought of it. He was very angry at the other answers."

greatest importance"–can be seen from his statement, in thanking Lindsey for "Paley's three volumes," that "Mr. Cooper has read the first volume, and says it is very valuable. How he will be impressed by the whole, is uncertain."[229] By September, however, Priestley was writing in tones of great acerbity about Cooper: "As to the *law*," he wrote of another prospective immigrant's hopes of living in America:

No person can practice till he has been in the country four years. This materially affects Mr. Cooper. I see he will make nothing out as a farmer. He has spent much money, as all say, very unwisely, tho he has got about as much by buying and selling land. However, all he has to look to is his profession as a lawyer, and for this he must wait a long time; till, I fear, he will be quite aground. I am very sorry that he is with us on more accounts than one.[230]

Priestley's feelings were apparently by November 1795 shared by his elder son. In November 1794 young Priestley had promised Watt that he would do all that was in his power for Cooper, although the "failing of the scheme" would necessarily make this more difficult. But, he wrote, "I esteem him much & am really sorry for him." A year later, however, he was writing in terms of some disillusion. He had, "to serve Cooper," taken up his lands. This had however not only not answered his own purpose: it had not materially altered Cooper's situation:

I am afraid that notwithstanding all I have done for him it will be of no service to him. He wants prudence and attention to small concerns. I have besides lent him money. More I cannot do & to see him as I think going back & unable to do him any good makes my situation very unpleasant.[231]

If it was Cooper's improvidence and lack of belief which made him less than wholly welcome company to the Priestleys at this time, there was perhaps another reason for their disillusion. For it was, as others also observed, Thomas Cooper's enthusiastic and in many ways misleading description of the lands in America which had materially contributed to the disastrous situation in which they found themselves – and about which their friends in England were so anxious. Priestley in December 1794 might write to one prospective emigrant that "after seeing Mr. Cooper's book, you cannot want any information that I can give you respecting this country; and knowing the facts, you must advise yourself. It is too hazardous for another to give it."[232] But from another prominent English emigrant there came a striking criticism of Cooper's role. "It strikes any man upon his first arrival in America," wrote Ralph Eddowes to William Roscoe,

---

[229] Priestley to Cooper, 6 April 1794; Priestley to Lindsey, 10 February 1795, *Works*, I.2.294–5.

[230] Priestley to Lindsey, 14 September 1795: D. W. L. Mss., for passage largely omitted in Rutt, *Works*, I.2.318.

[231] J. Priestley, Jr. to J. Watt, Jr., 20 November 1794, 10 November 1795, B. R. L.

[232] Priestley to G. Clark, 22 December 1794, *Works*, I.2.286.

that many of the descriptions given of it in England have been too highly coloured—much has been done in that way from interested motives in land-holders. Our acquaintance C—r is much to be blamed for having lent himself to persons of this sort and having given tempting accounts of plans and things which he never saw. I presume you have heard that the scheme of the new settlement in which he was engaged has fallen through.[233]

It was a criticism of which Cooper himself was clearly aware. He had heard, he wrote to Watt in his letter of July, "that People say Dr. Priestley & I are dissatisfied with the Country: I desire, & *the Dr. desires*, this may be contradicted, for I do not believe either he or I have felt that sentiment—for a moment. I have nothing to unsay about the Country, but," he significantly added, "I shall recommend it to no one any more."[234] In the autumn of 1795, as Priestley's letter indicates, Thomas Cooper seems to have decided to relinquish his attempt at farming, and turn instead to the law. "I have not stirred from this neighbourhood since I have been here," he wrote to Watt in October 1795, "not having yet from my farm proviso frugis in annum Copia. The price of everything has increased here *at least* a third since I was first in the Country." In the following spring he reported from Philadelphia his still straitened circumstances, and one remarkable adjustment which these had apparently forced upon him. The difficulty of finding servants was such that he was seeking for his wife "a set of negroes. . . . They are purchaseable here till they are 28 Years old." He lamented his failed land speculations, and described his decision to cease farming as a livelihood:

I soon found that altho' farming was profitable to a man who cd. steadily follow up his Servants & be among them, that it did not suit me otherwise than as a very healthy & very pleasant employment; I have therefore taken to the Law & am now here to be admitted into what we call our supreme Court; against all rule I have obtained the privilege of pleading tho' inadmissible for want of sufft. residence. . . . Another twelve months of uphill exertion,

Cooper optimistically concluded, "& then I think I shall feel a little at my ease."[235]

---

[233] R. Eddowes to William Roscoe, 4 February 1795, Roscoe Papers, 1331, L. P. L. For Eddowes, and his emigration to America in 1794, cf. his letters to Roscoe, in Roscoe Papers, L. P. L.; Geffen, *Philadelphia Unitarianism*, 37–8; and D. A. B. For similar comments on Cooper's misleading description of America, cf. Wansey's "Journal," 39; and also B. Waterhouse to Adams, 15 August 1799, Mass. Hist. Soc., Adams Papers, Reel 396: "Cooper's book proves him to be either a fool or knave; and Dr. P. must know that it is surcharged with falsehoods; untruths which have had a cruel operation on his countrymen who have been seduced by it to come purchase land here." For Cobbett's comments, cf. *Porcupine's Works*, XI.396, n.

[234] T. Cooper to Watt, Jr., 4 July 1795, B. R. L. Cooper was at this period, however, certainly not entirely persona non grata with the Priestley family: cf. T. Cooper to W. Russell, 19 April 1796: after staying in Philadelphia while Priestley was also there, he accompanied Priestley's daughter-in-law home—"Mrs. P. and I with the negro got safe home"—and sent his regards to "the Ladies and the Dr." (Russell Papers, Penn. Hist. Soc., Bundle XI.6.) And cf. J. Priestley, Jr. to J. Watt, Jr., 18 December 1796, B. R. L.

[235] Cooper to Watt, Jr., 5 October 1795, 4 April 1796, B. R. L.; and cf. also Cooper to (?T. and R. Walker), 29 September 1795, ibid.; above, n. 234, and Priestley to Wilkinson, 17 December 1795, W. P. L.: "now we only hire a black slave by the week."

From February to April 1796, Priestley was also in Philadelphia. The city whose preachers in 1794 had, to his intense annoyance, prevented him from preaching from their pulpits—and who were still intent on doing so—enthusiastically welcomed the "Discourses on the Evidences of Divine Revelation," which he delivered in the Universalist Church. "I have the use of Mr. Winchester's pulpit every morning," he wrote to Lindsey, "and yesterday preached my first sermon to a very numerous, respectable, and very attentive audience. I was told there was a great proportion of the members of Congress, though the notice of my preaching was very imperfect." There was, he was sure, "a good prospect of establishing an Unitarian congregation in this City"; he had, he wrote, "promised to officiate every winter without any salary, which I shall absolutely refuse, provided I can lodge with a friend, which, I fancy, will always be easy. . . . Mr. Russell thinks it very practicable. I propose to be here two months, and in that time shall feel my way better."[236] From the Unitarian congressman George Thatcher there came on the day of Priestley's first preaching in Philadelphia, a eulogistic account. "I have just returned from the Universal meeting house," he wrote to James Freeman in Boston,

and I hasten to tell you I had the pleasure of hearing our friend Dr. Priestley. He came to this City on Tuesday evening, and though it was but sparsely known, or hinted that it was probable he would preach this morning, the meeting house was very much crowded; and I believe I may safely add—he gave universal satisfaction, for as I returned in the street it seemed as if every tongue was engaged in speaking his praise, or answering the clergy of the City.[237]

"The congregations that attended were so numerous that the house could not contain them," recorded one who attended Priestley's preaching in Philadelphia: "so that as many were obliged to stand as sit, and even the door-ways were crowded with people."[238] By April Priestley could report to Lindsey that he had now preached six of his "Discourses,"

---

[236] Priestley to Lindsey, 15 February 1796, *Works*, I.2.333–4.

[237] Thatcher to Freeman, 14 February 1796, *P. M. H. S.*, Series 2, Vol. 3. (June 1886): 39–40.

[238] Priestley, *Works*, I.2.333, note. For Thatcher, a prominent Unitarian Congressman, cf. *D. A. B.* For the establishment of a Unitarian congregation in Philadelphia after this visit of Priestley's, in which he was assisted by Russell, John Vaughan, and two other prominent English radical emigrants, Ralph Eddowes, and Joseph Gales, cf. Priestley's letters to Lindsey, 11 September 1796, 13 January, 3 April 1797, *Works*, I.2.352–3, 369, 375; Geffen, i.e., *Philadelphia Unitarianism*, 32–55; and E. M. Wilbur, *History of Unitarianism in Transylvania, England, and America*, 396–7. For Eddowes, a wealthy merchant from Chester, who had been educated under Priestley at Warrington, cf. above, n. 233, and *Monthly Repository*, 9 (1814): 265. For Gales, cf. *D. A. B.*; Geffen, 42–4; W. H. G. Armytage, "The Editorial Experience of Joseph Gales, 1786–1794," *North Carolina Historical Review*, 28 (1951): 332–61; W. S. Powell, ed., "The Diary of Joseph Gales, 1794–1795," ibid., 26 (1949): 335–47; and W. G. Briggs, "Joseph Gales, Editor of Raleigh's First Newspaper," *North Carolina Booklet*, 7 (1907): 105–30. And cf. also Jefferson to Benjamin Waterhouse, 19 July 1822, Ford, ed., *Works of Jefferson*, XII.244: "When I lived in Philadelphia, there was a respectable congregation of (Unitarians) with a meeting-house and regular service which I attended, and in which Dr. Priestley officiated to numerous audiences." For its decline, cf. Geffen, 54–5.

and with more acceptance, if I may judge from appearances, than I could have imagined, the congregations having always been numerous, and most respectable; and it is evident that I am heard with more attention than was given to me in any place before. A considerable proportion of the members of Congress, and all the principal officers of state, are my constant hearers. As Mr. Adams, the vice-president, is most punctual in his attendance, and an old acquaintance and correspondent, I shall dedicate the discourses I am delivering to him.[239]

"He is admired and caressed by all classes of citizens," wrote Benjamin Rush of this most successful of Priestley's visits to Philadelphia: "by the democrats for his *political* and by the aristocrats for his *religious* principles."[240] And writing to Lindsey, Priestley made clear his enjoyment of his visit, not only for the success which he enjoyed in the pulpit, but for the companionship of his friend William Russell, in whose house he stayed, and with whom he visited the President—where they "spent two hours as in any private family. He invited me to come at any time, without ceremony," Priestley wrote. It was during this visit to Philadelphia that Priestley certainly encountered Talleyrand, who was elected a member of the American Philosophical Society on 15 April, at a meeting at which Priestley was present. Priestley saw much also, but for the last time, of Rittenhouse, visiting, as in 1794, with his family frequently, and among "a select few" who dined with Rittenhouse on 18 March 1796. Rittenhouse was also among those present at a dinner given for Priestley in the University of Pennsylvania. Priestley was on close terms with James Woodhouse, who had succeeded to the chair of chemistry, and with Drs. Ewing and Andrews, provost and vice-provost of the University.[241]

From Philadelphia in the spring of 1796 Priestley wrote one of his most enthusiastic endorsements of America: "Every thing here," he wrote to Lindsey,

---

[239] Priestley to Lindsey, 8 April 1796, *Works*, I.2.336; and cf. Priestley to Belsham, 5 March 1796, ibid., 334-5. For the dedication to Adams, cf. Priestley, *Discourses relating to the Evidences of Revealed Religion* (Philadelphia, 1796, 1797), *Works*, XVI.3-5; and also Adams, *Works*, I.488; Z. Haraszti, *John Adams and the Prophets of Progress*, 281. And for the great impression which Priestley's preaching made, cf. also Priestley to Wilkinson, n.d. (February/March 1796), and same to same, 28 July 1796, W. P. L.; and also Rush to John Dickinson, 5 April 1796, L. H. Butterfield, ed., *The Letters of Benjamin Rush*, II.773-4. For a characteristically dissenting voice, and an unvarnished account of the offence which Priestley gave to some of his hearers in one Discourse, in his description of the practices of the ancients, cf. T. Cooper to Watt, Jr., 4 April 1796, B. R. L. (although cf. also Rush to Griffith Evans, 4 March 1796, *Letters*, II.773).

[240] Rush to Griffith Evans, 4 March 1796. For the wording of this sentence, in which "democrats" was transcribed as "autocrats," cf. Butterfield's note. The present writer has followed his suggested emendation, which accords entirely with Priestley's recorded political conversations in Philadelphia (cf. above, n. 168 and below, n. 259). Cf. also Rush to John Dickinson, 5 April 1796, for his great admiration for Priestley: "I have never met with so much knowledge accompanied with so much simplicity of manners."

[241] Priestley to Lindsey, 15 February 1796, *Works*, I.2.332; Earl, "Talleyrand in Philadelphia," 294 and note; E. Ford, *David Rittenhouse, Astronomer-Patriot, 1732-1796* (Philadelphia, 1946), 196-7, 206-7; Brooke Hindle, *David Rittenhouse*, 348, 361; W. Barton, *Memoirs of Rittenhouse*, 438; D. J. Boorstin, *The Lost World of Thomas Jefferson* (Boston, 1948), 17; Priestley, *Works*, I.2.341-4; and D. A. B. For the invitation from the Philosophical Society for Priestley to be their president on the death of Rittenhouse, in June 1796, which Priestley declined, cf. Priestley to B. S. Barton, 8 October 1796, Schofield, *Scientific Autobiography*, 291-2: "on many accounts you ought to prefer Mr. Jefferson." And see below, n. 262.

is the reverse of what it is with you. Indeed, I do not suppose there ever was any country in the world in a more flourishing and promising way. I wish all my friends, with you, were here, provided they could subsist and be happy. But great numbers find themselves, on one account or other, disappointed, and return, I understand, with very unfavourable ideas of the country; and for this I see no remedy. I have been careful not to encourage any person to emigrate, though I admire this country very much.[242]

It was an admiration which he could at this time extend apparently without reserve for the country's political processes, although the measures under discussion in Congress were by now the cause of great party passion, and were the result of policies of which he certainly did not approve. "Mr. Jay's Treaty is almost universally condemned," he had written to Lindsey in August 1795, "so that many think the President will not ratify it."[243] In Philadelphia in the following spring, however, he was a witness to the impassioned debates which carried the Treaty into effect: "After a long discussion," he reported to Lindsey,

the House of Representatives have voted, by a majority of three, for carrying the treaty with England into execution. Having much leisure, I have attended to hear much of the debate, and have heard as good speaking as in your House of Commons, and much more decorum. A Mr. Amos (sic: Ames) speaks as well as Mr. Burke; but, in general, the speakers are more argumentative, and less rhetorical. And whereas there are not with you more than ten or a dozen tolerable speakers, here every member is capable of speaking, which makes interesting debates tedious.

He was, he added, "well acquainted" with the members of both parties— "and they do not avoid one another, as the heads of parties do in England; and when once anything is decided by fair voting, all contention ceases."[244]

"We are about to change our President," Priestley wrote to William Vaughan in November 1796: "But, tho the contest will be a very warm one, it will be attended with no serious inconvenience." To Lindsey, however, he wrote of the increasing tension between France and America: "But amidst so much calamity affecting Europe, it would be extraordinary indeed if we could wholly escape." Washington, he commented, was "much blamed by the zealous republicans for his ingratitude to France, and, I think, with some reason." But, he added, "I steer clear of all politics, and indeed feel very little interest of any thing of that kind here. I feel as an

---

[242] Priestley to Lindsey, 15 February 1796, *Works*, I.2.332-3. And cf. ibid., XVI.500-11, for Priestley's *Discourse* to the Emigrant Society in Philadelphia in 1797.

[243] Priestley to Lindsey, 12 August 1795, ibid., I.2.315.

[244] Priestley to Lindsey, 3 May 1796, ibid., I.2.340. And cf. Page Smith, *John Adams*, (New York, 1962), II.892, Priestley's attendance at the debates, and his admiration for the oratory of the Federalist, Fisher Ames. (Adams to Abigail Adams, 28 April 1796.) For Priestley's certain disapproval of Jay's Treaty, however, which he was later to write "could not fail to give umbrage to France," and should not have been made without her approval, cf. Priestley, *Letters to the Inhabitants of Northumberland* (2d. ed., 1801), *Works*, XXV.168: "In this proceeding I see nothing of the fairness and openness that I should have expected from a republican government." (ibid.). For the debates, cf. also Elkins and McKitrick, 447-9.

Englishman, and shall sincerely lament any evil that may befal my native country, though I condemn as ever the conduct of its rulers."[245]

With the retirement of Washington from the Presidency of America in the winter of 1796-7, the affairs both of that country, and of all Europe, were approaching a state of great crisis—as Priestley's correspondence was increasingly to reflect. To the republicans of America, the ratification of Jay's Treaty with England had represented a betrayal of their country's true interests, an unseemly dependence upon England, and an unnecessary affront to republican France. "The Treaty is generally disliked," Thomas Cooper had also reported in September 1795, "and Jay has been repeatedly burnt in effigy."[246] The acquiescence in British rights of search on the high seas; the surrender of American trading rights in the Caribbean and Canada, while allowing the British full entry to all American ports and passage up the Mississippi, was, wrote Jefferson, "an infamous act, which is nothing more than a treaty of alliance between England and the Anglomen of this country against the legislature and people of the United States."[247] Jefferson's enthusiasm throughout this period for the cause of France, and his conviction that for republican principles to be established firmly in America, they must also be established in England, was tempered only by his recognition of the frequently aggressive and incursive policies of France in America itself. His overriding fear of English influence, and its effect upon the commercial and financial policies of the Administration, was, however, very frequently expressed. "The aspect of our politics is wonderfully changed since you left us," he wrote in his notorious letter to Philip Mazzei in the spring of 1796, shortly after the ratification of the Treaty with England:

In place of that noble love of liberty, & republican government which carried us triumphantly thro' the war, an Anglican monarchical, & aristocratical party has sprung up, whose avowed object is to draw over us the substance, as they have already done the forms, of the British government. The main body of our citizens, however,

he added,

remain true to their republican principles; the whole landed interest is republican and so is a great mass of talents. Against us are the Executive, the Judiciary, two out of three branches of the legislature, all the officers of the government, all who want to be officers, all timid men who prefer the calm of despotism to the boisterous sea of liberty, British merchants and Americans trading on British capitals, speculators and holders in the banks and public funds, a contrivance

---

[245] Priestley to William Vaughan, 4 November 1796, *Scientific Correspondence*, 152; Priestley to Lindsey, 3 December 1796, *Works*, I.2.363.

[246] T. Cooper to (?T. and R. Walker), 29 September 1795, B. R. L.: although, as he added, French machinations in America had threatened to turn the tide of opinion the other way.

[247] Jefferson to Rutledge, 30 November 1795, *Works*, VIII.199-201; and cf. Peterson, *Jefferson and the New Nation*, 545-52; Banning, *The Jeffersonian Persuasion*, 234-8, for the opposition to the Treaty.

invented for the purposes of corruption, and for assimilating us in all things to the rotten as well as the sound parts of the British model.[248]

Jefferson's own deeply held convictions as to the proper course to be followed in the economic, social and political development of America had been most perfectly expressed in his *Notes on Virginia*, written in 1781-2, privately distributed among his philosophical acquaintance in 1785, and finally published in 1787. In these Jefferson had written of his conviction that, unlike the countries of Europe, America should establish herself as an agrarian nation, using to the full her abundant resources, and relying for her imports upon the manufactures from across the Atlantic.[249] Jefferson was, however, even as he distributed a strictly limited edition of the *Notes* to his friends,[250] serving as the American Minister Plenipotentiary in Paris, with the specific mission of negotiating treaties of commerce with the nations of Europe in conjunction with Franklin and John Adams. To John Jay he privately wrote at this time that, "were we perfectly free to decide this question," he would encourage his countrymen only to cultivate the land:

However we are not free to decide this question on principles of theory only. Our people are decided in the opinion that it is necessary for us to take a share in the occupation of the ocean, and their established habits induce them to require that the sea be kept open to them, and that that line of policy be pursued which will render the use of that element as great as possible to them. I think it a duty in those entrusted with the administration of their affairs to conform themselves to the decided choice of their constituents: and that therefore we should in every instance preserve an equality of right to them in the transportation of commodities, in the right of fishing, and in the other uses of the sea.[251]

Jefferson, as Hamilton was later to declare, was a "theorist who never allowed his dogmas to interfere with the practical exigencies of public affairs."[252] But in the early 1790s when America, it seemed, was in danger of a radical departure from his ideal, Jefferson, with the assistance of many able propagandists, among them the Quaker Congressman George Logan, was moved to restate his fundamental philosophy as an alternative to the economic policies of Hamilton, and the spirit of corruption, speculation and degrading dependence upon England that

---

[248] Jefferson to Philip Mazzei, 24 April 1796, Ford, ed., *Works*, VIII.235–41.

[249] Jefferson, *Notes on the State of Virginia*, M. D. Peterson, ed., *Thomas Jefferson* (New York, 1984), 290–1; and cf. also ibid., 300–2.

[250] Cf. Jefferson's presentation inscription in his copy to Price: Boyd, ed., *Jefferson Papers*, VIII.246; and also Price to Jefferson, 2 July 1785, ibid., VIII.258–9; Peterson, *Jefferson and the New Nation*, 248.

[251] Jefferson to John Jay, 23 August 1785; and cf. also to Hogendorp, 13 October 1785: "You ask what I think on the expediency of encouraging our states to be commercial? Were I to indulge my own theory, I should wish them to practice neither commerce nor navigation, but to stand with respect to Europe precisely on the footing of China. We should thus avoid wars, and all our citizens would be husbandmen . . . But this is theory only, and a theory which the servants of America are not at liberty to follow." (Boyd, ed., *Jefferson Papers*, VIII.426–8, 631–4).

[252] Cit. C. A. Beard, *Economic Origins of Jeffersonian Democracy* (New York, 1913), 427–8.

they entailed. Even at this time, Jefferson did not repudiate the legitimate role of commerce. It was, nevertheless, in the 1790s that he above all articulated his agrarian beliefs as a matter of public policy in opposition to the direction in which Hamilton's policies seemed to be leading America. The agrarian interest, wrote Jefferson, was vital for the survival of America: "Such men are the true representatives of the great American interest, and are alone to be relied on for expressing the proper American sentiments."[253] His deep distrust of the policies of the Federalists led to his continued impassioned outpourings against the English. They had, he wrote in the spring of 1797, "wished a monopoly of commerce and influence with us; and they have in fact obtained it." As a result of the overriding influence which the English had established in the affairs of America, they could, he believed, now force their former colonies "to proceed in whatever direction they dictate, and bend the interests of this country entirely to the will of another." It was, as a result,

impossible for us to say we stand on independent ground, impossible for a free mind not to see and to groan under the bondage in which it is bound. If anything after this could excite surprise, it would be that they have been able so far to throw dust in the eyes of our own citizens, as to fix on those who wish merely to recover self-government the charge of subserving one foreign influence, because they resist submission to another.

And Jefferson wrote of his fear that "after plunging us in all the broils of the European nations, there would remain but one act to close our tragedy, that is, to break up our Union."[254]

The effects of Jay's Treaty on both the internal and external policy of America – arousing a storm of anti-British sentiment, and a deepening of French hostility towards America – had, by March 1797, when Jefferson arrived in Philadelphia to assume the Vice-Presidency, become sufficiently marked for John Adams to deem it of the first importance to "avert a rupture with that nation, a rupture which would convulse the attachments of this country," and to propose the "necessity of an immediate mission to the Directory." For the power of France at this time was formidable; and her ambition no less so. In December 1796 only a violent gale off the coast of Ireland had prevented her invasion of that country; she was in possession of the navies of both Spain and Holland; nor were her territorial aims limited to Europe. Her anger at the provisions of Jay's

253 Jefferson to Colonel Arthur Campbell, 1 September 1797, Ford, ed., *Works*, VIII.336-8. And cf. Beard, *Jeffersonian Democracy*, 415-35; and also Banning, *Jeffersonian Persuasion*, 204-5. Recent controversy on the Jeffersonian philosophy of the 1790s has, as Banning has written, resulted in such disparate interpretations as to lead to the inevitable puzzlement of general readers, and "an imposing barrier to further study" for historians: L. Banning, "Jeffersonian Ideology Revisited: Liberal and Classical Ideas in the New American Republic," *W. M. Q.*, 3rd Series, 43 (1986): 3-4; and for a summary of the debate, cf. J. Ashworth, "The Jeffersonians: Classical Republicans or Liberal Capitalists?" *Journal of American Studies*, 18 (1984): 423-35. For an account which demonstrates the degree to which Jefferson and his followers did attempt to accommodate the commercial ideal to the agrarian, cf. D. McCoy, *The Elusive Republic, Political Economy in Jeffersonian America* (Univ. of N. Carolina Press, 1980). For Logan, cf. *D. A. B.*
254 Jefferson to Elbridge Gerry, 13 May 1797, Ford, ed., *Works*, VIII.283-8.

Treaty had extended to a rescinding of her formal alliance with America, the recall of her ambassador to Paris, and a systematic policy of depredation of American shipping in the Caribbean and the Atlantic. At the same time, the arrogance of France, the lessons which her militarism and perversion of the revolutionary ideal seemed to carry, and the scarcely veiled appeals now being made by her retiring Ambassador, Adet, to the people of America in the columns of the Philadelphia *Aurora*, were strengthening the hand of those who, as Jefferson and many others believed, wished to strengthen the monarchical at the expence of the republican features of the American Constitution. In April the news reached America of the expulsion by the Directory of the new American ambassador to Paris, and of the humiliating and threatening terms in which the final interview of the retiring ambassador, Monroe, with the Directory had been couched. On 15 May a special session of Congress was called, and Adams on 16 May addressed both Houses in belligerent terms. He deplored the treatment of Pinckney, and the tendency of the French President's speech to Monroe—its "disposition to separate the people of the United States from their government; to persuade them that they have different affections, principles, and interests, from those of their fellow-citizens, whom they themselves have chosen to manage their common concerns; and thus to produce divisions fatal to our peace. Such attempts," said Adams, "ought to be repelled." And he authorised, while expressing his sincere desire for peace, an augmentation of the American naval establishment, to protect her now threatened commerce, and to act as "the natural defence" of the country. The militia, also, must be reorganized, armed, and disciplined. America, he declared, should not be involved in the political system of Europe, but it was "necessary, in order to (sic) the discovery of the efforts made to draw us into the vortex, in season to make preparations against them." In June 1797, three envoys from America, Elbridge Gerry, Pinckney, and Marshall, were despatched to France in order to restore the shattered state of Franco-American relations.[255]

In the spring of 1797, as the new Administration was starting out on its term of office, with considerable differences between Adams and Jefferson on the very threatening situation now facing America,[256] Priestley was once more in the city for an extended stay after the untimely death of his wife. "I expect to be there in December," he wrote to Lindsey, "and shall stay till the rising of Congress."[257] Shortly after

[255] N. E. Cunningham, Jr., *The Jeffersonian Republicans: The Formation of Party Organisation, 1789-1801* (Chapel Hill, 1957), 101ff.; Banning, *The Jeffersonian Persuasion*, 251; F. B. Sawvel, ed., *The Complete Anas of Thomas Jefferson* (New York, 1903); Ford, ed., *Works of Jefferson*, VIII.235-294; Chinard, *Jefferson*, 324ff.; Malone, *Jefferson and the Ordeal of Liberty*, 312-15; Adams, *Works*, IX.111-17; Elkins and McKitrick, 520-1, 550-3; Tagg, *Aurora*, 293-4.

[256] F. B. Sawvel, ed., *The Complete Anas of Thomas Jefferson*, 184-5, 2 March 1797; Jefferson to Madison, 1, 30 January 1797, Ford, ed., *Works*, VIII.262-4; 279-80; Chinard, *Jefferson*, 321ff.

[257] Priestley to Lindsey, 29 October 1796, *Works*, I.2.359-60; and cf. Priestley to Lindsey, 19 September 1796, ibid., I.2.354: "I never stood in more need of friendship than I do now. . . . This day I bury my wife." And also Priestley to Belsham, 8 October 1796, ibid., I.2.357; Priestley to William Vaughan, 1 November 1796, *Scientific Correspondence*, 151-2; and Priestley to Wilkinson, 19 September 1796, W. P. L.

his arrival he could write in optimistic terms of the outcome of the presidential election. "I seldom trouble you about the politics of this country," he wrote,

Indeed, I think very little about them. But I must inform you that Mr. Adams is to be our next President, and Mr. Jefferson our Vice-President, and there is no doubt they will act very harmoniously together, which will greatly abate the animosity of both parties. But such is the temper and habit of this country,

he could still confidently write, "that if anything be once decided, though by a single fair vote, all contention instantly ceases, and all will join with the majority."[258] With John Adams in January he was his usual forthright and, in the circumstances, far from politic self: "Dr. Priestley breakfasted with me," Adams wrote to his wife: "I asked him whether it was his opinion that the French would ultimately establish a republican government. He said it was." "He was," Adams later wrote, "very sociable, very learned and eloquent on the subject of the French revolution. It was opening a new era in the world and presenting a near view of the millenium (sic)."[259] This, however, appears to have been the last occasion on which Adams and Priestley enjoyed even the appearance of a friendly relationship. For Adams absented himself conspicuously from the second set of "Discourses" which Priestley was now delivering in Philadelphia; and in this he was not to be alone. "Partly from the novelty of the thing being done away," his son later wrote, "partly from the prejudices that began to be excited against him on account of his supposed political opinions (for high-toned politics began then to prevail in the fashionable circles) . . . they were but thinly attended in comparison to his former set."[260] "I suppose," Priestley later wrote of Adams, ". . . he was not pleased that I did not adopt his dislike of the French." And the increasing impact of party feeling can be seen in his very different comment on the politics of America to Belsham, only two months after his optimistic report to Lindsey: "We have got a new presidency," he wrote, "and I hope a more promising one than the last, though it will

---

[258] Priestley to Lindsey, 13 January 1797, Works, I.2.370–1.

[259] Adams to Abigail Adams, 26 January 1797, C. F. Adams, ed., Letters of John Adams addressed to his Wife (Boston, 1841), II.241–2: "and his opinion was founded upon the prophecy," Adams quoted Priestley as saying: "France appeared to him to be one of the horns that were to fall off. . . . When statesmen found their judgments upon prophecies," Adams concluded, "I can never confide in their opinions." Adams recounted this conversation with Priestley many years later to Jefferson: Adams to Jefferson, 15 August 1823, L. J. Cappon, ed., The Adams-Jefferson Letters, II.594–5. Adams then wrote that it was "not long after the denouement of the tragedy of Louis 16th," and while he was Vice-President, that "my friend the Dr. came to breakfast with me alone." Citing only this letter, Clarke Garrett, "Joseph Priestley, the Millennium, and the French Revolution," Journal of the History of Ideas, 34 (1973): 51–2, and also Respectable Folly, 133, dates the conversation with Priestley to 1794: but Priestley did not see Adams in 1794: cf. above, n. 142. And the exactly similar description of the actual conversation which Adams gave in the letter of 1797 suggests that this was the occasion to which he was referring in this later letter to Jefferson. Adams was still Vice-President in January 1797.

[260] Priestley, Memoirs, I.193–4; Priestley to Belsham, 11 January 1798, Works, I.2.391.

be difficult, and I fear impossible, to undo the false steps that have been taken."[261]

By the time that he was writing in such a strikingly different vein, Priestley had been — for the first time — much in the company of Jefferson, who had arrived in Philadelphia from his retreat at Monticello early in March. "I have seen a good deal of him," Priestley wrote. And if a turning point in his attitude towards American politics can be detected, it can be dated to this visit to Philadelphia, when he first became personally acquainted with this most articulate and influential of the "Anti-Federalists" — as, Benjamin Rush recorded, he was greatly wishing to do.[262] Only a year later Priestley was to contribute, in the Philadelphia *Aurora*, his *Maxims of Political Arithmetic*, which expounded the developing Jeffersonian social, political, and economic position in response to the financial and commercial policies of the Federalists. As Priestley's first piece of published political comment since his arrival in America some four years earlier, the *Maxims* represented a significant shift in the focus of his views from those which he had held for the very different social and economic circumstances prevailing in England. As such they arguably bear testimony not only to the influence which Jefferson surely had upon his thinking, but also to his increasing familiarity and identification with the social and political problems of what, in spite of all his protests to the contrary, was increasingly becoming his adopted country.

Priestley's instinctive sympathies, as his correspondence from 1794 onwards reveals, were with the Francophile policies of the Anti-Federalists, and as such hardly represented any alteration in the views which he had for so long, and amidst so much unpopularity, held in England. But in commenting upon the domestic issues at stake between the parties in America, in endorsing the Jeffersonian distrust of the particular emphasis which was accorded to the mercantile interest by the Federalists, his *Maxims* represented an adjustment of his views in accordance with what he clearly believed to be the true interests of the developing American social and political economy. In articulating these ideas, which were to play an influential role in Jeffersonian propaganda, Priestley's four years of experience in the backwater of rural Northumberland, and his first-hand knowledge of the need for the development of the vast hinterland of the American continent, were perhaps as crucial as was his encounter with Jefferson in the capital.

---

[261] Priestley to Belsham, 11 January 1798; same to same, 14 March 1797, ibid., I.2.373.

[262] Ibid; Rush to Jefferson, 4 February 1797, *Letters*, II.786. Jefferson had left Philadelphia on 5 January 1794, and did not return from Monticello until 2 March 1797. On 3 March he was installed as President of the American Philosophical Society, to which he had been elected on the death of Rittenhouse. On 4 March he was inaugurated as Vice-President of the United States. (Malone, *Jefferson and the Ordeal of Liberty*, 161–8, 295–7, 340–1; Peterson, *Jefferson and the New Nation*, 517, 560–3; Ford, ed., *Works*, VIII.136–7, 271; G. Chinard, "Jefferson and the American Philosophical Society," *Proc. Am. Phil. Soc.*, 87 [1944]: 267.) For an account of the meeting on 10 March at the Philosophical Society, at which Jefferson was flanked by Priestley and Volney, see Peterson, *Jefferson*, 576.

PRIESTLEY'S BREACH WITH THE FEDERALISTS
AND COBBETT'S ATTACK
1797–1799

As political tensions reached great heights in America, Priestley was not to be able to resist the excitement of the debate, and contribute his own pronounced views upon it. He was at this time closely associated with those circles of French and radical opinion in Philadelphia which were to be particularly the target of Federalist abuse. And he was also, largely on account of his personal and financial responsibilities (but also from his sense of his increasing unpopularity in the prevailing political climate in America) anxious to visit France. As early as December 1796 it was announced in the *Moniteur* that he intended settling there. And in a letter to Wilkinson, in January 1797, he attempted to explain how such an apparently erroneous report could have originated. "The conversations I had with the French Ambassador, the Bishop of Autun, and other French people, have led them, I suppose, though without any just grounds, to think that I should go to France."[263] Some two months later, however, while he was still in Philadelphia, Priestley heard the news from England of his son-in-law's bankruptcy, and of his daughter's distress, and almost immediately he wrote to Lindsey, "I am about to go to France, as I see that my property in the French funds will never yield me any thing while I remain here, especially as this country is now on bad terms with France." And he wrote of his plight in an unguarded note to Adams, asking, "as a friend," for some help:

Now as I presume that you will soon send messengers or dispatches to France, could you favour me with a passage thither? In return, it might be in my power to render some service to this country with persons of influence in that; and this I should be happy in taking every opportunity of doing.[264]

Early in April, although apparently without any response from Adams, his plans were definite: "I believe I shall go with the late French ambassador, M. Adet," he wrote to Lindsey,

and Mr. Lister (sic: Liston), the English minister, will give me a protection in case of meeting with an English ship of war. He does the same for M. Adet; so that a better opportunity I could not have had. If I succeed, I shall make some pur-

---

263 Priestley to Wilkinson, 25 January 1797. Cf. also *Réimpression de l'Ancien Moniteur* (Paris, 1858–63), 21 November 1796, and Chaloner, 38–9.
264 Priestley to Lindsey, 3 April 1797, *Works*, I.2.375; Priestley to Adams, n.d. (March/April 1797), Mass. Hist. Soc., Adams Papers, Reel 383, and Appendix.

chase of land in France, and then I can spend my time here or there, as it shall suit me.[265]

On 11 April he wrote to Wilkinson that he had just heard from Joseph: "He is much against my going to France, but I do not think I can do better, and hope to go with M. Adet, tho, as he is not now at Philadelphia, the matter is not yet settled."[266]

This plan, however, came to nothing. Writing to Lindsey on the eve of his departure for Northumberland from Philadelphia, however, Priestley could not conceal his uneasiness at his situation now in America, as well as his mounting concern for the state of England:

I hope and pray that a kind Providence may watch over you and my other friends in the great crisis, and, if prudent measures be taken to prevent tumults, the calamity may not be so very great as we have sometimes apprehended. It is impossible, however, not to be exceedingly anxious about the issue when so much is depending.

"The shock given to credit," he continued, referring to the dramatic suspension of payments by the Bank of England in February 1797, "affects this country in a very sensible manner; which, joined with our unpleasant situation with respect to France, fills the country with alarm. The Congress will soon meet, but what they will do is very uncertain. I am sorry," he continued, "to see a dislike to France prevail so generally as it does. This affects me and all who are supposed to wish well to that country." And he made one of his first complaints about Cobbett, whose *Porcupine's Gazette,* its columns dedicated to the calumniation of all "Gallo-American patriots," and encouragement of the Federalists' policies towards France, began its career in Philadelphia in the spring of 1797:

The writer of that scurrilous pamphlet on my emigration now publishes a daily paper, in which he frequently introduces my name in the most opprobrious manner, though I never took the least notice of him; and have had nothing to do with the politics of the country; and he has more encouragement than any other writer in this country. He, every day, advertizes his pamphlet against me, and after my name adds, "commonly known by the name of the fire-brand philosopher."

---

[265] Priestley to Lindsey, 3 April 1797, *Works,* I.2.375; and cf. also to Wilkinson, 1 April 1797, W. P. L.: "I shall probably sail in about a fortnight. . . . Perhaps I may spend much of my time there too, though I feel the strongest attachment to the place where my wife lies, and wish to be buried near her." Priestley, as he wrote to Wilkinson in January 1797, believed that he would never "completely recover the state of mind" he had before the death of his wife; and his "unsettled" state, as he described it also to Lindsey, undoubtedly contributed to his eagerness to go to France to try to realise his assets in the French funds. (Priestley to Wilkinson, 25 January 1797, W. P. L.; Priestley to Lindsey, 3 April 1797, *Works,* I.2.378.) For Adet, a distinguished chemist, who entered into controversy with Priestley, cf. F. J. Turner, ed., *Annual Report of the American Historical Association* (1903), II.728n.; 1003–9; Schofield, 293. And cf. Cunningham, *The Jeffersonian Republicans,* 101; and *The Gros Mousqueton Diplomatique; or Diplomatic Blunderbus, containing Citizen Adet's Notes to the Secretary of State . . . With a Preface by Peter Porcupine,* November 1796.
[266] Priestley to Wilkinson, 11 April 1797, W. P. L.

"The aversion to those emigrants from England," Priestley continued, "who are supposed to have been hostile to the measures of government there, is greater, I think, than it was in England. But," he added, still apparently with little sense of the extremes to which party spirit was to lead the political scene in America, "happily, we are better protected by the laws, and the disposition of the lower orders of the people, among whom a respect for the French, for assisting them in gaining their liberty, is not extinguished. The rich," he wrote, however, in tones which certainly echoed the prevailing Jeffersonian hostility towards the Federalists, "not only wish for alliance offensive and defensive with England, but, I am persuaded, would have little objection to the former dependence upon it."[267]

It was when the political scene in America was beginning to show dangerous signs of the extremes of party spirit, but before he had fully realised the extent to which this might rebound upon him, that Priestley made one more overture to the politician whom he had once counted as his friend, and whom he still clearly believed would not allow the prevailing controversies to interfere with their friendship. It was, however, an overture which was to cost Priestley, and the friend for whom he made it—Thomas Cooper—dearly. For Priestley, in asking Adams to consider Cooper for a government office—that of Agent for the American claims—was to be decisively ignored by Adams, after he had, moreover, admitted much unity of political sentiment between Cooper and himself. "It is true," wrote Priestley,

that both Mr. Cooper and myself fall, in the language of calumny, under the appellation of *democrats*, who are represented as enemies to what is called *government* both in England, and here. What I have done to deserve this character you well know, and Mr. Cooper has done very little more. In fact, we have both been persecuted for being the friends of liberty, and our preference of the government of this country has brought us both hither. However, were the accusation in any measure true, I think the appointment of a man of unquestionable ability, and fidelity to his trust, for which I will make myself responsible, will be such a mark of superiority to popular prejudice, as I should expect from you, and therefore I think it no unfavourable circumstance in the recommendation.[268]

Subject now to open attacks from Cobbett, and his relationship with Adams clearly deteriorating, Priestley remained in Northumberland throughout the summer and autumn of 1797, anxiously aware of the uncertain political outlook, and increasingly dispirited about his own situation.[269] In September 1797 he reported to Lindsey that there had been a meeting of "our *College* or rather *Academy*," but this was one of his last references to what was to prove an abortive venture, of which, as

---

[267] Priestley to Lindsey, 30 April 1797, *Works*, I.2.377–8.
[268] Priestley to Adams, 11 August 1797, Mass. Hist. Soc., Adams Papers, Reel 385. And cf. Malone, *Cooper*, 87–9; Smith, *Freedom's Fetters*, 174.
[269] Priestley to Lindsey, 29 May, 18 June 1797, *Works*, I.2.380–2; Priestley to Wilkinson, 7 September 1797, W. P. L.

his son was later to write, "little more was done" during his father's life-time "than to raise the shell of a convenient building."[270] In September 1797 Priestley wrote to Wilkinson that he had effectively abandoned further hope of preaching in Philadelphia: "Philadelphia is so disagree-able a place, that I shall hardly go thither any more. I have done all I could with respect to my real object. The novelty of my preaching is over, and, with that, its effects." "The proposal for my *Church History* gets me only seven subscribers in Philadelphia," he shortly afterwards wrote to Ben-jamin Vaughan, "and among them was not Mr. Adams, tho' he received in a flattering manner the dedication of a volume of my Discourses."[271]

In the summer of 1797, Benjamin Vaughan, after a delay of more than a year, had finally arrived in America. During his sojourn in France in 1796 he lived at the house of the American consul, Fulwar Skipwith; he saw much of the Stone brothers; and he renewed his friendship with Tal-leyrand. Benjamin Vaughan almost certainly at this time produced a polit-ical pamphlet, *De l'Etat Politique et Economique de la France sous la Consti-tution de l'An III*, which urged support for the French constitution under the Directory. And he arrived in America carrying letters of introduction from his fellow emigré, Thomas Paine, testifying to their "private friend-ship," and acknowledging great confidence in Vaughan's "public princi-ples." To Priestley's lasting regret, however, Benjamin Vaughan chose on his arrival to retreat into anonymity on the family estates in Maine.[272] His attempts to persuade Priestley to visit him there failed, as had Russell's similar endeavours to persuade him to travel to Boston. "You are very kind in endeavoring to facilitate my journey to Kennebec," Priestley wrote to Benjamin Vaughan in April 1798, "but at my age, which you seem to overlook, but the effects of which I feel, it is too formidable an undertaking. Crossing the Atlantic appears much less formidable."[273]

---

[270] Priestley to Lindsey, 14 September 1797, D. W. L. Mss., letter not in Rutt; Priestley, *Memoirs*, I.169–70; and below, n. 318.

[271] Priestley to Wilkinson, 7 September 1797, W. P. L.; Priestley to Benjamin Vaughan, 19 April 1798, *Scientific Correspondence*, 153.

[272] For Priestley's still eager expectations of the arrival of Benjamin Vaughan, cf. Priestley to Lindsey, 28 July 1796, passage omitted in Rutt: "We hear nothing of *Mr. B. Vaughan*, tho we expect him dayley." And also Priestley to William Vaughan, 1 November 1796, *Scientific Correspondence*, 152: "I wonder we hear nothing of your brother." For Benjamin Vaughan's eventual arrival cf. Priestley to Lindsey, 18 June 1797: passage omitted in Rutt: "B. Vaughan, I hear, is arrived in this country. He spent a day with Mr. Russell on his way to Boston. I hardly expect to see him, or Mr. Russell, any more." Benjamin Vaughan's arrival in America is frequently stated to have been in 1796 (cf. *D. A. B.*). But Priestley's letter makes it clear that it was in 1797. Cf. also Samuel Vaughan, Sr. to Benjamin Vaughan, 8 November 1797: "My dear Benj I have received yours of August 29 which gave me great pleasure after so long a Silence as it advises me of your arrival." (A. P. S., Vaughan Papers, B. V. 46 p). See also Murray, *Benjamin Vaughan*, 381; Marvin, *Benjamin Vaughan*, 55–60. For an account of Vaughan's time in France cf. Murray, 361–73; Marvin, 35–54. For the authorship of the pamphlet, see especially Murray, 362, note. For two lengthy memoranda which Vaughan also wrote on English politics, in October 1794 and October 1795, see also ibid., 352–6.

[273] Priestley to Benjamin Vaughan, 19 April 1798; and also Priestley to Hurford Stone, 20 January 1798, *Works*, I.2.393–4: "You congratulate me on my interview with B. V. He is settled so far from me (at Kennebech) that I never expect to see him at all. I once intended to have gone as far as Boston; but travelling in this country is so inconvenient and expen-sive, that I have given up all thoughts of it. I should almost as readily cross the Atlantic."

"As the day draws to a close with me," he wrote to his former pupil, "I have no time to lose."[274] But in addition to his scientific and philosophical pursuits, which he pursued with his usual unremitting industry in Northumberland, Priestley was now undoubtedly more willing to be drawn publicly into the politics of America. He saw at this time, as he later wrote, "almost all the newspapers that are printed in Philadelphia," of which above all others he gave the preference to Bache's *Aurora*, the increasingly outspoken mouthpiece of francophile and Anti-Federalist opinion. Not even the *Morning Chronicle* in England was, Priestley believed, "superior to the *Aurora* with respect to just sentiment, valuable information and good composition." It had, he wrote, materially helped "the people in general" to be "better informed concerning their true interest, and their real friends." It would appear to have been in the summer of 1797 that he first attended the local festivities for the Fourth of July. And his inclinations, as his son wrote of him at this time, to give vent to political sentiments freely in conversation, were clearly to the fore by the autumn of that year. "In the present extraordinary situation of things in the political world, the universal topic of all conversation is of course *politics*," he wrote to Lindsey in November, "though less in this remote part of the country than in cities."[275] He continued, however, to take a close interest in the politics of Europe. He was convinced, as he wrote on more than one occasion to Lindsey and to others, that the present war would only end with "the total destruction of the European monarchies." And he welcomed, in the late autumn of that year, the declaration of the Directory in France before the renewed outbreak of hostilities, having, as he wrote, "some faint hopes that it may prevent them." He continued to hope that his fellow countrymen would finally be induced to make peace. "Surely, having all the power of France to contend with *alone*, must create some alarm in the most confident," he wrote; and, on hearing of the victory over the Dutch at Camperdown: "This peace, so long looked for, must surely come soon. I wish the victory at sea, on the coast of Holland, of which we have just heard, may lead to it; but I rather fear it may put it farther off." "We are entering on another year," he wrote on 1 January 1798,

which seems to be big with great events. May they be happy ones! But I cannot help fearing great calamity, as the prophecies announcing such, I think are about to be accomplished, or rather are accomplishing. If so, they must ruin us all (tho there may be interruptions, and alleviations of distress) in the downfall of all the monarchies of Europe.

He took, he continued to assure Lindsey, little interest in the affairs of America:

nothing here much interests myself, I only feel as an Englishman, and sincerely wish the peace and happiness of my native country: Tho I have little to complain

---

[274] Priestley to Benjamin Vaughan, 19 April 1798.

[275] Priestley, *Letters to the Inhabitants of Northumberland*, *Works*, XXV.112, 128, note, 129; Priestley, *Memoirs*, I.200; Priestley to Lindsey, 16 November 1797, *Works*, I.2.387, and D. W. L. Mss.

of here, I feel like a banished man; tho the banishment respects my friends more than the country. But I hope we shall all meet in a better country, and one much better governed.[276]

It was in 1798, however, shortly after making these elaborate disclaimers of political involvement, that Priestley was fatally to compromise himself—and was to be fatally compromised—in more than one way. He attempted to undertake once more the journey to France which he was later so strenuously to deny; he expressed himself in an unguarded fashion in an astonishingly frank political correspondence with one of Philadelphia's leading Federalists, the Unitarian Congressman George Thatcher; and he was to be struck a terrible blow with the publication, in London, of letters written to him and Benjamin Vaughan from John Hurford Stone in Paris—with their publication in London in May, and subsequently, in August, by Cobbett in Philadelphia. And if he was to exercise little true judgment in his response to this disaster, the contribution which he made at the beginning of the year, in February 1798, to Bache's *Aurora*, must cast doubt on all his previous protestations of lack of thought for American affairs. For the Essay which he then published, under the pseudonym "A Quaker in Politics," his *Maxims of Political Arithmetic*,[277] was an attack upon the Federalist policy of equipping America with a navy as a protection against the aggression of France, and in particular her depredations on American commerce. It also forcefully advocated the emerging Jeffersonian counter-philosophy of the need, in the present state of the development of America, to give priority to her internal needs and industry, to avoid the entanglements of European politics, and to refrain from giving to the commercial interest the protection of what must necessarily be a powerful and expensive naval force. Priestley also commented, in a characteristically provocative fashion, on the extremes to which he feared the present party animosities might lead the country.

"Without inquiring into the cause," Priestley began this indictment of Federalist policies,

. . . it is a fact, that the conveyance of goods, or the carrying trade of this country, which has generally been taken up by the merchants, though it is no necessary branch of their business, is peculiarly hazardous, and, of course, expensive. This expense the country at large must pay, in the advanced price of the goods purchased. In this state of things they have also found it necessary to send ambassadors to distant countries, in order to remove the supposed cause of the difficulty, which is attended with another expense. It has likewise been thought necessary to build ships of war for the purpose of protecting this carrying trade; and if this be done to any effect, it must be attended with much more expense.

I do not pretend to be able to calculate the expense occasioned by any of these circumstances; but the amount of all three, viz. the additional price to the carrier for his risk, the expense of ambassadors, and that of fitting out ships of war, I

---

[276] Priestley to Lindsey, 4, 16, 30 November 1797; 1 January, 17 May 1798, *Works*, I.2.385–90; 399–401; and D. W. L. Mss. for passages omitted in Rutt.

[277] *Aurora*, 26, 27 February 1798.

cannot help thinking must be much more than all the profit that can be derived from the carrying trade; and if so, a person who had the absolute command of all the shipping, and all the capital of the country, would see it to be his interest to lay up his ships for the present, and make some other use of his capital. And as the greatest part of the country is as yet uncleared, and there is a great want of roads, bridges, and canals, the use of which would sufficiently repay him for any sums laid out upon them, and they would not fail to contribute to the improvement of the country, which I suppose to be his estate, he would naturally lay out his superfluous capital on these great objects. The expense of building one man of war would suffice to make a bridge over a river of a considerable extent, and (which ought to be a serious consideration) the morals of labourers are much better preserved than those of seamen, and especially those of soldiers.

The solution, Priestley wrote, was for the merchants to operate entirely at their own risk; for ambassadors to be everywhere withdrawn, and for the resources of America to be applied not to such potentially warmongering activities to which commerce inevitably led, but to the arts of peace: to the purchase of books, and philosophical apparatus—"of which all the universities and colleges of this country are most disgracefully destitute," so that "men of letters coming to reside here find their hands tied up"; and to much greater ease of travel.

It was perhaps, Priestley wrote, "the wise plan of Providence" to involve America "in the vortex of European politics, and the misery of European wars, and to prevent the importation of the means of knowledge till a better use would be made of them. Nations make slower advances in wisdom than individual men, in some proportion to their longer duration." But he wondered at the height that party animosity had reached—"when all were so lately united in their contest with a common enemy." And he speculated that the divisions within the country between those who favoured France, or England, were now such that "the decision of the government in favour of either of them" could "be the cause of a civil war. But even this, the most calamitous of all events, would promote a greater agitation of men's minds, and be a more effectual check to the progress of luxury, vice, and folly, than any other mode of discipline. . . . Many lives, no doubt," he added,

will be lost in war, civil or foreign; but men must die; and if the destruction of one generation be the means of producing another which shall be wiser and better, the good will exceed the evil, great as it may be, and greatly to be deplored, as all evils ought to be.[278]

---

[278] Priestley, *Maxims of Political Arithmetic, applied to the Case of the United States of America, Works*, XXV.177–81. In all Priestley's extant correspondence there appears to be a remarkable absence of reference of any kind as to the motives and timing of the composition of the *Maxims*. That the idea for them originated in the conversations which Priestley had in Philadelphia in 1797, there seems to be little doubt. On the other hand, it was not until the spring of 1800 that Priestley sent them, when they were published as an Appendix to his *Letters to the Inhabitants of Northumberland*, to an appreciative Jefferson (below, n. 368). For their almost certain influence upon Thomas Cooper, which has been generally neglected by historians, cf. below, n. 312. The importance which Priestley now placed upon the "improvement of the country," was possibly a new departure from the particular

Maxims of Political Arithmetic was Priestley's first contribution to political debate since the second part of his *Appeal to the Public* was published
in 1792. It was certainly no less controversial, and its publication undoubtedly signalled an awareness on Priestley's part of the problems facing
America, an adaptation of his views to the needs of his adopted country,
and a willingness to offer his own characteristically radical contribution.
Writing to Benjamin Vaughan shortly afterwards, however, Priestley
sharply replied to a suggestion which his former pupil had apparently
made: "I cannot imagine what you mean by intimating that some *event*
may take place here in which I may be of some use. Whatever it be, you
certainly overrate my importance in fancying this. All that I can possibly
do here is in *theology*, and that is over." "I wish," he added, however, "*you*
had been nearer the seat of government you might have been of some
use in the present awkward state of things, tho' I know there is a great
jealousy of the interference of foreigners, I keep out of the way of all Politics," he further and rather inaccurately stated. And he lamented again
the abuse to which he was subject:

and in a newspaper most patronized by the governing people. But these things
do not affect me much now. I had hoped, however, that, while I was quiet myself,
I might have been quiet in this country; to peace I have always been a friend.
But the jealous friends of the revolution here are in general out of favour now,
and the tories are courted and popular.[279]

In the spring of 1798 Priestley was undoubtedly aware of the violent
party spirit raging in America, and, although he remained aloof from the
actual strife, not averse to expressing his own very extreme opinions
upon it. "Our governors, in my opinion, have acted as absurdly as
yours," he wrote to Lindsey, "and have brought the country into great
difficulties, which might easily have been avoided, and the measures
they are now taking have no tendency to make things better."[280] He read
with consuming interest also, the English papers which his friend sent
him, the *Morning Chronicle* and the *Cambridge Intelligencer*: "I value them
much, especially the Cambridge paper," he wrote. Aware, as he repeatedly

---

emphasis of his own previous published thinking on the subject (above, pp. 12–13). On
the other hand, he had in his *Lectures* of 1788 followed Adam Smith in recognising the importance of agriculture as the basis of any economy. As Durey, in "Thomas Paine's Apostles,"
679–80, rightly points out, Priestley was not in the *Maxims* arguing against commerce as
such. And as McCoy has also stated, Priestley's position in America (as it was to be echoed,
the present writer would suggest, by Thomas Cooper) could be said to be akin to that of
Smith's, on the necessary allocation of resources in an underdeveloped economy (below,
n. 312).

[279] Priestley to B. Vaughan, 19 April 1798, *Scientific Correspondence*, 153–4. For Priestley's
continuing regret, throughout 1798, at the distance separating him from effective communication with Benjamin Vaughan, cf. Priestley to J. Vaughan, 21 January, 27 September 1798:
"I am very sorry that I cannot have an interview with your brother, and indeed very seldom
hear from him, but we are too far asunder for the purpose." (A. P. S., Priestley Papers,
B. P. 931.)

[280] Priestley to Lindsey, 17 May 1798, *Works*, I.2.400.

wrote, of the unpopularity which his political views had brought upon him — by the summer of 1798, indeed, professing that he did not have a friend in Philadelphia who was willing to receive him there[281] — and encouraged by a letter which he received apparently early in the year from Hurford Stone in Paris, he was once more seriously considering moving to France. "What will you say to my leaving this country and going to France," he asked Lindsey, in March:

Because, it is not very unlikely to take place, and I really think I shall not *get*, nor *do*, much more good here. At the solicitation, I believe, of M. Adet, the last ambassador from France to this country, the French directory have made a decree in my favour, allowing me 1200 livres (About 50 Sterling) per an. but my interest in their funds shall amount to as much. This is certainly very little, considering how much was deposited with them for my use.

"But," he added, "M. Delacroix, the minister for foreign affairs, who transmitted the decree to the French consul here, tells me that, if I would go to reside in France, my losses would be more than made up to me. In this case I think that, all things considered, I ought to go." It would, undoubtedly, be a loss to leave his library and apparatus, now "in excellent order," but the account from Hurford Stone of the progress of rational Christianity in France had given him great encouragement. "Having," he concluded, "some acquaintance with Talleyrand Perigord, late Bishop of Autun, who succeeds M. Delacroix, I have written to him, almost promising, that if what his predecessor said would be made good, I should go. . . . I wait his answer."[282]

To Hurford Stone, from whom he said he had heard with pleasure, and whose regular correspondence he regretted, Priestley wrote that "the last thing" Talleyrand had said to him before his own departure from America, "was, that he expected to see me in France." He described to his friend the "most extraordinary change in the politics of the trading people of this country since I came hither, as to countries in alliance with France, which gives me great concern." He wrote of the abuse to which he was subjected, but, he said, he remained convinced that "the bulk of the people are still hostile to England, and rejoice, as I do, in the success of the French, and I am persuaded would never be brought to fight against them. I hope," he added, "that you have more moderation and good sense than to proceed to hostilities against this country, though it has not (I mean the leaders of it) deserved any better. It is the mercantile interest only that has made the change, and the glorious success of

---

281 Priestley to Thatcher, 26 July 1798, *P. M. H. S.*, Series 2, Vol. 3 (June 1886): 23: "Mr. Vaughan expressed, I doubt not, a sincere wish to do it, but said he had not convenience."

282 Priestley to Lindsey, 8 March 1798, *Works*, I.2.397; and D. W. L. Mss., for the large part of this letter omitted by Rutt. Cf. also Priestley to Wilkinson, 21 January, 15 March 1798, W. P. L.; and also T. Cooper to J. Watt, Jr., May 1798, B. R. L.: "The Dr. whose Judgement is extremely weak in my opinion (entre nous) is inclined again to visit England. . . . To me, the loss of their Society would be a heavy one." Cf. also Chaloner, 29.

French arms, if nothing else, will open their eyes at length."[283] To Belsham he wrote of the great superiority of the friends of France "to their opponents, and so they are in Congress."[284]

His own views and sympathies in little doubt in his correspondence with his closest English friends, it was nevertheless to a confirmed Federalist, the Unitarian Congressman George Thatcher—whose own political prejudices were so pronounced at this time that he openly accused the editor of the *Aurora* of being an agent for the French—that Priestley in America confided most freely, and in what was to prove to be damaging detail, in the spring and summer of 1798. By the summer of that year, war hysteria was sweeping America. As the French indulged in fresh conquests in Europe, and in the invasion of Switzerland and the brutal suppression of revolt in the Batavian republic alienated many of their last remaining friends; as they assembled on the northern coast of France a great armament with the openly avowed purpose of the invasion of England (and in England itself further undeniable treason was uncovered against the government) the Directory in Paris treated those American envoys who had been despatched to repair the relationship between France and America with undisguised contempt—refusing to receive or negotiate with them until the Americans had agreed to pay a large sum of money, and appealing over their heads and that of the American government to "the French party in America." "The American people," said Talleyrand, in a letter published on 16 June to widespread indignation and outrage, by Bache in the *Aurora*,

will not commit a mistake concerning the prejudices with which it has been desired to inspire them against an allied people, nor concerning the engagements which it seems to be wished to make them contract to the detriment of an alliance, which so powerfully contributed to place them in the rank of nations, and to support them in it.

The chicanery, duplicity, and barely concealed ambition of the French in these attempts at negotiations produced, as Jefferson on first reading the despatches of the envoys described it, "such a shock on the republican mind, as has never been seen since our independence." The Directory's behaviour, he wrote, was "calculated to excite disgust and indignation in Americans generally, and alienation in the republicans particularly." Jefferson's own reaction was modified by his censures on Adams's policy towards France. But from all over America, on the news of the results of the negotiations and the return of the envoys, came an explosion of rage against all things French. An outpouring of publications demonstrated the perfidy of France not only now, but even when

---

[283] Priestley to Hurford Stone, 20 January 1798, *Works*, I.2.393–4: "M. Adet promised to write to me on his arrival in France, but I have not heard from him since he left us. . . . I wrote to M. Perigaux to desire he would make a small purchase for me near Paris, but my money in his hands will not suffice for that purpose. P.S. Perhaps M. Talleyrand would assist in what I have hinted. . . ."

[284] Priestley to Belsham, 11 January 1798, *Works*, I.2.391.

apparently America's true friend, during the War of Independence; and an effective witch-hunt was begun against all those suspected of belonging to the French network of spies and infiltrators which it was believed, not without some justification, had permeated the country. Carrying its measures with little difficulty through Congress, the Administration "carried the United States into a virtual state of undeclared war with France." A Navy Department of State was established; the building of several more ships for the navy authorised; the army considerably augmented, and Washington summoned from his retirement at Mount Vernon to command it. From all quarters patriotic addresses poured in upon the President, to which Adams replied in terms reflecting the prevailing emotion: "There is nothing in the conduct of our enemies," he declared to the Inhabitants of Burlington, New Jersey, "more remarkable than their total contempt of the people, while they pretend to do all for the people; and of all real republican governments, while they screen themselves under some of their names and forms. . . .

The American people are unquestionably the best qualified of any great nation in the world, by their character, habits, and all other circumstances, for a real republican government; yet the American people are represented as in opposition, in enmity, and on the point of hostility against the government of their own institution and the administration of their own choice. If this were true, what would be the consequence? Nothing more nor less than that they are ripe for a military despotism, under the domination of a foreign power. It is to me no wonder that American blood boils at these ideas.

As a climax to the war fever now prevailing, the Adams Administration passed through Congress in June and July 1798 the Alien and Sedition Acts. A foreigner suspected of being "dangerous to the peace and safety of the United States" could now be deported by Presidential decree, without the right of redress or appeal; and any writing deemed defamatory of the government was punishable by substantial fines and imprisonment. The Acts led to the precipitate departure of more than one distinguished French emigrant, to the prevention of the arrival of others, and to the prosecution of several of the more outspoken opponents of the Administration, including Bache, of the *Aurora*. Their provisions had however alarmed even Hamilton: "Let us not establish a tyranny," he wrote, on reading the "highly exceptionable" provisions of the Sedition Bill. The Acts satisfied the demands of the more extreme Federalists, and curtailed the activities of some of their opponents. But they aroused others to a fresh pitch of indignation. They led directly to the first organised expression of the doctrine of States' rights against abuses of them by the Federal Constitution, in the Virginia and Kentucky Resolutions; and they clearly defined the issues on which the Jeffersonians were to oppose Adams in 1800. They were laws passed "so palpably in the teeth of the Constitution as to shew they mean to pay no respect to it," wrote Jefferson.

The most long-sighted politician could not, seven years ago, have imagined that the people of this wide-extended country could have been enveloped in such delusion, and made so much afraid of themselves and their own power, as to surrender it spontaneously to those who are manoeuvring them into a form of government, the principal branches of which may be beyond their control.

The government of America had now, Jefferson believed, "become more arbitrary, and has swallowed more of the public liberty than even that of England."

In the correspondence which Jefferson and Adams exchanged many years later, both men warmly defended the positions they had taken on the Alien and Sedition Acts. "We were then at War with France," Adams wrote:

French Spies then swarmed in our Cities and in the Country. Some of them were, intollerably (sic), turbulent, impudent and seditious. To check these was the design of this law. Was there ever a Government, which had not Authority to defend itself against Spies in its own Bosom? Spies of an Ennemy (sic) at War?

To which Jefferson, who had suffered much at the hands of Federalist obloquy, replied, recalling vividly "the sensations excited in free yet firm minds, by the terrorism of the day. None can conceive who did not witness them, and they were felt by one party only."[285]

Against this background Priestley, from March 1798 onwards, carried on his correspondence with Thatcher. He wrote of his continuing conviction that "the present war will not end without the downfall of all the European monarchies, that of England . . . included"; of his concern, at the same time, at the latest news from France, for "though a well wisher to the cause of that country, I shall very sensibly feel any injury done to England, or to America"; and of his hopes that "your resolves in Congress will be temperate." His friends in England were, he was sure, too complacent about the possibility of a French invasion; and, he added, "whether there be peace or war, there must be a revolution in that country." Of the situation facing America, he also voiced his opinion: "The French," he wrote,

whose successes, like those of the Romans, appear to have made them, as they were at the time of Jugurtha, equally void of fear or shame, want to bully you out of a sum of money, but I do not think they will seriously go to war with you. I should think that suspending all intercourse with them (if) they knew themselves better, would best answer your purpose. They cannot hurt you here, and if the merchants will trust their property at sea, let it be at their own risk, and not involve the country. The consumer will pay, and much less so, and more equally, than by any tax for defence. This conduct of the French does not,

he added, "affect the Constitution, which does not differ essentially from that of this country nor the people at large. It is only the character of the

[285] Smith, Freedom's Fetters, 435–42; Adams, Works, IX.182ff.; A. Beveridge, The Life of John Marshall (New York, 1916), II.214–373; Ford, ed., Works of Jefferson, VIII; Cappon, ed., Adams-Jefferson Letters, II.329–32; A. Koch and H. Ammon, "The Virginia and Kentucky Resolutions: An Episode in Jefferson's and Madison's Defence of Civil Liberties," W. M. Q., Series 3, Vol. 5 (1948): 145–176; Peterson, Jefferson and the New Nation, 590ff.; Banning, The Jeffersonian Persuasion, 251ff; Elkins and McKitrick, 555–79, 581–93; Tagg, 342–3, 377–88.

people now in office who may change tomorrow; as it is to be hoped they will soon. But if I meddle with your *Politicks*," he concluded, "I shall be more abused by P. Porcupine than I am."[286]

In May 1798, clearly in response to some riposte from Thatcher, Priestley assured him that "the unanimity you speak of in this country does not exist in this neighbourhood. The gentry, indeed, are generally with you, but the lower classes those, who must take the field, had rather fight the English than the French. They do not," he provocatively added, "so soon change their sentiments and habits as their superiors." Later in the month he revealed that he appreciated something of the danger of speaking his mind so freely. He would, he hoped, see Thatcher during the next winter, "if your new *alien bill* do not confine me, or send me out of the country. I have always been reckoned a dangerous man." "It does not become an *alien* to say much about Politicks, especially in these dangerous times," he wrote. But he could not resist a diatribe against the increasingly warlike measures of Adams, and his undisguised hostility towards the French:

I should think that, much as you may approve the measures of your President, you must begin to think that his language is too close a copy of that of Mr. Pitt, and even of Peter Porcupine. It is not *statesmanlike*, not, I think, prudent even in a state of open hostility, much less during a negociation for peace. If ever there be a restoration of harmony, his abusive language must be retracted, or suppressed. But I fear that the irritation must now be so great, that there cannot be any good understanding between this country and France while he is President. . . . What can Mr. Adams mean by calling the French liberty *chimerical*. What then is that of America? The two constitutions do not differ in anything essential.[287]

In June and July 1798, however, the Adams Administration passed the Alien and Sedition Acts, and Priestley on 5 July was concerned to make a disclaimer of any great interest in politics: "I only read the Newspapers once a week, and seldom anything more than the articles of news. I have not even read any of your debates in Congress in all this session. But on this account," he again provocatively added, "I may view the subject with more coolness, and perhaps in a truer light." He hoped, however, that Thatcher would not take what he had written in jest, in earnest, "or let any thing I write go beyond yourself. For the times, I perceive, grow venous, and a man's (sic) whose thoughts only do not go with the current may be in danger." To his not unjustified dismay, however, he discovered by the end of that month that his "'democratic principles,'" as expressed in these private letters, had been well publicised in Philadelphia— "'which in the present season,'" as his informant wrote, "'have irritated some persons not a little.' Now," wrote Priestley,

I have not written to any person in Philadelphia besides yourself, and I am sure you would not *intentionally* expose me to danger. However, I will take care

---

[286] Priestley to Thatcher, 10 March, 10 May, 19 April 1798, *P. M. H. S.*, Series 2, Vol. 3 (June 1886): 18–20.

[287] Priestley to Thatcher, 10, 31 May 1798, ibid., 20–21.

to *send no more, lest a worse thing come unto me.* I find I am at the mercy of one man, who, if he pleases, may, even without giving me a hearing, or a minutes warning, either confine me, or send me out of the country. This is not a pleasant situation.[288]

It was, however, a situation which Cobbett, seizing his opportunity, was mercilessly to exploit. In May 1798, letters from Hurford Stone in Paris to both Priestley and Benjamin Vaughan in America had been captured on board a Danish vessel, and published in London.[289] In their free admission of his correspondents' knowledge of and acquaintance with the politics and chief actors on the political scene in France, and in their open approval and, indeed, undisguised satisfaction, at the progress which that country was making in the subjugation of Europe, these letters were to compromise the reputation of Priestley for many, almost irreparably, both in England and, after their publication by Cobbett, in America. As a result of the obloquy to which they exposed him, moreover, Priestley was to be led by degrees into an explanation of his political principles which upset further some of his closest friends, and plunged him into the minefield of political controversy in America.

It was, Hurford Stone wrote in his letter to Priestley, "now a very considerable time" since he had had the pleasure of hearing from Priestley himself. News about him continued to reach France, however, and his friends there were delighted that he was contemplating, on the restoration of peace, a visit to Europe, and possibly even a permanent settlement in France. "Whether you fix yourself here or in England, *(as England will then be),"* he wrote, in a sentence which was to be much fastened upon, "is probably a matter of little importance."[290] On this point, however, Priestley had "now a friend on the Continent who can discuss this . . . with you better than myself."[291] And Stone's letter—"something like a

---

[288] Priestley to Thatcher, 5, 26 July 1798, ibid., 23–4: "I apprehend," he added, "my pleading the zeal that, in conjunction with Dr. Price, I always shewed for the liberty and independence of America, would not avail me now. None of my friends of similar principles can now join me here. They would not, I presume, be allowed to land in the country." And cf. Priestley to Lindsey, 2 August 1798, D. W. L. Mss., passage omitted in Rutt: "Tho I take no part in Politicks, I am regarded with much suspicion. I sent Mr. Belsham a specimen of the light in which I am considered here. It will amuse you. But, as I was told lately, even Dr. Price, if he were alive, would not be permitted to come hither. I am now liable to be confined, or sent out of the country, without a minute's warning, or having a hearing given me."

[289] (J. H. Stone), *Copies of Original Letters recently written by Persons in Paris to Dr. Priestley in America. Taken on board of a neutral Vessel* (2nd ed., London, 1798): Preface: London, May 14, 1798: "The Letters of which the following are literal copies were found on board of a Danish ship, lately brought into one of our ports, by the Diamond Frigate. The originals were inclosed in a cover directed to 'Dr. Priestley, in America.' They have been exhibited, with the usual attestations, in the high court of admiralty, as part of the evidence in the proceedings against the above-mentioned ship, and her cargo, and are now remaining on record in the public registry of that court. Their authenticity is, therefore, placed beyond a dispute, and may be personally ascertained by any man who chuses to take that trouble."

[290] Ibid., 11–12.

[291] Ibid., 12; the reference is surely to Benjamin Vaughan.

packet," as he rightly described it—was in effect a lengthy apologia for the politics of the Directory in 1798, both of their aims towards England, and their expansionist policies throughout Europe. Of the benefits of this brand of republican freedom, Stone was not in the least doubt. "You will of course have heard," he wrote,

that our OLD COUNTRY is now the only one left to struggle against the French Republic, and left *under every disadvantage that every friend to her real welfare would wish;* namely, in a very fair way of accomplishing your prophetic discourses, delivered at various times, and divers manners, of which *happily* they took no account.

You will have heard of the vast armaments and preparations of every kind which have been making for some months past, and which are carrying forward with all that energy and activity which characterizes this nation, when they have a purpose in hand which they must go through, cost what it will. . . . The invasion of England is a *denrée*, or merchandize of the first necessity for them, and I should doubt whether any concession on the part of England could now avert the experiment: whether it will be a fatal one to it's government, time only can determine. In the mean time, the government here are putting in work every engine, attempting to engage every passion, to enlist every prejudice, nevertheless always anxious to discriminate between the Government and the People, flattering the one, as much as they profess to execrate the other.[292]

"Whatever can tend to humble the English government is most anxiously sought after, in whatever shape the mode of opposition presents itself," Stone added, in the course of his long review of the politics of Europe. "You have heard," he wrote,

of the destruction of the government of Venice, of the regeneration of that of Genoa, of the constitutional fermentation of the Cisalpine Republic;—the news of the present period is the fall of the Papal power, the possession of Rome by the French troops on account of the late massacre, and the formation of this country into a new government under the name of the Roman Republic. In like manner as the French troops are now employed in pulling down the chief spiritual power in one part, another portion is occupied in overturning the genius of Aristocracy in the Swiss Cantons, each of which, under the influence of the French Republic, are busied in destroying their present tyrannic oligarchies, and melting the whole into an Helvetic Republic, founded on the basis of the Rights of Man, with a representative government.

"The spirit of Equality," wrote Stone, in this enthusiastic eulogy of French military adventurism, "which has retraversed the Alps, has also entered the Rhine." He admitted, after discussing at some length the politics of Germany, and the probable fall of Spain, that "amidst these changes without," there were some jarring notes: "You will have trembled for our Constitution, and probably felt *some alarm for liberty* on the events of the 18 Fructidor; you will have felt similar *disagreeable sensations*, in hearing of the late arrests of the Deputies in Holland." These events were, he admitted, "no doubt, *very distressing*," but to be justified by events. The government of Holland, "though indebted for its political existence to

---

[292] Ibid., 12–14; 29.

France, has all along shewn a *most misplaced spirit of independence.*" And he proceeded to give an idyllic view of the present tranquil and civilised state of France. "The country, so far as respects its domicile state, is more advantageously situated than any other in Europe. Agriculture was never so much the rage, and manufactures, *but for the great encouragement given to English produce,* would have been equally thriving." France, he believed, would be "the residence of vast numbers at the peace. . . . The spoils of Italy are on their way to Paris," which in every respect could be regarded as the cultural capital of Europe. As to America,

Whether we shall continue or increase our hostilities towards the United States, is as yet uncertain; all depends on the great operation directing against England. If that succeeds, English influence will probably not predominate amongst you. In the mean time, it is most likely that the French will go on as at present, treating *with as little ceremony as usual* every thing that relates to America. John Adams's speech on the opening of congress caused a few smiles; the more so, as it was understood to be a speech full of thunder and menace against France. Nothing is wanting but the interposition of some upright and patriotic citizen, to settle the misunderstanding; but I fear *it will not be done in John Adams's time.*[293]

A more damaging publication than Hurford Stone's letter to Priestley in the circumstances in which it appeared in America, it would be hard to imagine. The letter accompanying it, moreover, written to one merely described, by Stone, as "M. B. P.," for delivery by Priestley, added further, if that were possible, to the suspicions as to the intentions of the latter in America, and of the connections which he had there.[294] For this letter, intended for Benjamin Vaughan, as Priestley was shortly to have no hesitation in avowing, not only drew, as in the letter to Priestley, a pleasing picture of the latest advances of the French in Europe. It specifically mentioned Stone's acquaintance with Talleyrand, and of the honour which he felt in his attentions, and added that "he continually enquires for you, and begs his best remembrances." It again justified the coup of 18 Fructidor; wrote approvingly of events in Switzerland, of a possible revolution in Spain, and of the "immense preparations" now being made for "the English expedition." Stone described also "the manufactory of which you laid the corner stone," as being now "finished," and forming "one of the finest establishments in France." Talleyrand, he added, was among the subscribers. His letter to Benjamin Vaughan concluded with a postscript, written certainly by his close friend, Helen Maria Williams, asking her correspondent if he was continuing his "speculations on the great events"; wondering if he was "in the press"; and mentioning "the books" which he had written. She informed him that she would send him "the first opportunity, the French translation of my Swiss Travels — for I have no English copy in my possession. . . . I flatter

---

[293] Ibid., 15–27.
[294] Ibid., 28: "I inclose a note for our friend M. B. P.; but as ignorant of the name he bears at present among you, I must beg you to seal and address it. We have heard nothing of him since his departure, and know but vaguely that he is secreted at Kennebeck."

myself," she wrote, of this encomium of French dominion in Europe in 1798—which Stone had recommended in his letter to Priestley—"you will approve the spirit in which it is written."[295]

In his *Gazette* on 20 August, under the dramatic title "PRIESTLEY COMPLETELY DETECTED," Cobbett devoted his entire two middle pages to a full publication of these letters, with the editorial comments from the English edition, and further vitriolic additions of his own. Priestley, he wrote, stood now fully revealed as a traitor, as he had always declared. His opinions, and those of his friends, were a danger to America, and Adams must deport him:

If this discovery passes unnoticed by the government, it will operate as the greatest encouragement that its enemies have ever received; they will say, and justly too, that though the President is armed with power, he is afraid to make use of it, and that the Alien-Law is a mere bug-bear.

Priestley's clear acquaintance with one who was on intimate terms with the leading men in France, his knowledge of this spy's secret hideaway, made it clear that

they look upon Priestley as one of their agents here, as acquainted with the spy, and with his whole business. . . . This explains the old Tartuffe's *intimacy with Adet*, and his being one of the party at the farewell festival given to that insolent agent of insurrection.

Confident as he was, Cobbett added, "that deep plots are going on in these states, I long ago said that French spies were everywhere at work." "*Vigilance*," he concluded, "ought to be the order of the day, and we see no vigilance any where. The evil will come on us by-and-by like a thief in the night, and I repeat my fears, that it will find us unprepared."[296]

Cobbett's publication of Hurford Stone's letters, and their subsequent appearance "in all the public papers" in America, "except those notoriously devoted to the cause of France,"—convincing some certainly that "Dr. Priestley is a French agent, & corresponds with their other Spies in this Country"[297]—so shortly after the passage of the Alien and Sedition

---

[295] *Copies of Original Letters*, 29–35; L. D. Woodward, *Hélène-Maria Williams et ses Amis*, 134–5; H. M. Williams, *A Tour in Switzerland, or, a View of the present State of the Government and Manners of these Cantons, with comparative Sketches of the Present State of Paris* (London, 1798); Murray, *Benjamin Vaughan*, 416–17; and cf. Stone to Priestley, *Copies of Original Letters*, 16.

[296] *Porcupine's Gazette*, 20 August 1798. J. M. Smith, in *Freedom's Fetters*, the most definitive history of the Alien and Sedition Acts, makes no mention of this journalistic coup by Cobbett against Priestley, which was to have widespread publicity, and undoubtedly added to the clamour that he be deported. Priestley was frequently to cite Cobbett's campaign against him as very influential among the Federalists (below, nn. 335, 338). This omission of evidence, tending as it does to suggest that Priestley was a figure of no political significance, at a time when Thomas Cooper, as Smith demonstrates in much detail, undeniably was, has further contributed to the general underestimate of Priestley's political standing in America. Cf. John Vaughan's comment, below, n. 336.

[297] Cobbett, *Remarks on the Explanation lately published by Dr. Priestley* (cf. below, n. 302): and *Porcupine's Works*, IX.247. Charles Nisbet to Charles Wallace, 25 October 1798, Miscellaneous Papers—Charles Nisbet, Rare Books and Manuscripts Division, The New York Public Library, Astor, Lenox and Tilden Foundations.

Acts, was a blow which Priestley could little have expected. "Your correspondence is one of the few consolations I have left," he wrote in a desolate letter to Lindsey: "Since your last the intercepted letters have been published here, with the preface and notes from the English edition, and others much more virulent." He quoted an extract of Cobbett, and added:

In this low manner am I continually treated, tho I have no more to do with the politics of the country than you have, and hardly think about them. So violent, however, is party spirit in this country and so general the prejudice against me as a *friend of France*, that if there be not a change soon I cannot expect to live in peace here.[298]

For Cobbett's paper, however, he almost immediately composed a reply. John Hurford Stone, he said, was a friend of many years' standing; a member of his congregation at Hackney; and "a zealous friend of the American and French Revolutions, which sufficiently accounts for his corresponding with me." He could not, however, he asserted, be held "answerable for what he or any other person may think proper to write to me." The letter enclosed for him was, he freely declared, written for Benjamin Vaughan—"formerly a pupil of mine, and son to Mr. Samuel Vaughan, who some time ago resided in Philadelphia." He was "a man that any country may be proud to possess; having for ability, knowledge of almost every kind, and the most approved integrity, very few equals." He was an acquaintance of the President, "who will smile at the surmises that have been thrown out on the subject." And he had "fixed his residence at Kennebeck, because his family has large property there. If he or I had been a spy in the interest of France," Priestley concluded, "we have made a very strange choice of situations in which to do mischief."[299]

From Benjamin Vaughan, in reply to a letter expressing concern and disbelief from his brother Charles—"S(tone) is a dangerous friend and your prudence would never have placed yourself in his power"—came a similar defence against the unwanted publicity that had been thrust upon him, and an avowal of his determination to retreat from all politics.

I can only say that Mr. S. had no invitation to write to me upon politics; & that . . . no letter or information whatever of any kind has passed from me to any one person on the continent of Europe, directly or indirectly, since I have been on the continent of America.

. . . Since politics have become warm here, I have in the same proportion avoided them; & my political acts have been none; my writings have been none; and my conversations highly guarded.

He had, he claimed, "for some time renounced all politics, from the deep persuasion, that providence has special objects in view in the world at this time, which are of the most extensive & important nature": but with

---

[298] Priestley to Lindsey, 6 September 1798, D. W. L. Mss.; cf. Rutt, *Works*, I.2.407, where there are omissions.

[299] Priestley to Cobbett, 4 September 1798, *Works*, I.2.406-7.

which he as an individual had little to do: "the issue is to be waited for by persons like myself, with the most profound and tranquil submission."[300]

In the aftermath of the furor created by Cobbett's publication of the intercepted letters, Priestley also wrote to John Vaughan. His son, he informed Vaughan, was intending to sail for England before the winter, "and if, in consequence of arrangements with his uncle, he settle in France, I shall meet him there in the spring." He would, he realised, lose much by leaving Northumberland, but he was now reconciled to the idea. To Lindsey in November he wrote similarly: "My son is preparing for his voyage . . . my future destination will depend on the interview between him and his uncle. If, in consequence of it, he should remove to Europe, I shall not stay behind; and indeed my continuance here appears less and less desirable."[301] In the event, however, Joseph Priestley, Junior sailed for England alone in January 1799 – to be pursued even there by the invectives of Cobbett. He stayed there for a full year, returning to America in the summer of 1800, and yet another of Priestley's schemes for leaving to settle in France was to come to nothing.[302] Instead, in the prolonged absence of Joseph, and while party animosities in America reached fresh heights of virulence, he found himself forced to reply to the calumny with which he was beset.

Already in December 1798 a note of defiance had entered Priestley's

---

[300] C. Vaughan to B. Vaughan, 19 September 1798, B. Vaughan to C. Vaughan, 25 September 1798, A. P. S., Vaughan Papers, B. V. 46 p, and Appendix; and cf. also B. Vaughan to C. Vaughan, 2 October 1798, ibid., cit. Murray, 421-2; and ibid., 421, for the furor in Boston, to which Benjamin Vaughan steadfastly refused to make any public reply.

[301] Priestley to J. Vaughan, 27 September 1798, A. P. S., Priestley Papers, B. P. 931; Priestley to Lindsey, 1 November 1798, D. W. L. Mss.: passage omitted in *Works*, I.2.409–11. It continues: "Here I hope I can be *quiet*, tho I am an object of much suspicion and dislike to all the friends of government, and to a degree that is really extraordinary, considering that I am well understood to have been a zealous friend to their revolution, and have not concerned myself with the Politicks of the country at all, as the course of my studies and publications will show."

[302] Cf. Priestley to Lindsey, 23 December 1798, D. W. L., passage omitted in *Works*, I.2.412: "This will be delivered to you, I hope, by my son. How happy should I think myself to be with him at the same time, tho but for an hour." Priestley to Lindsey, 21 March 1799, ibid., passage omitted in Rutt, I.2.415-16: "My son is, I hope, by this time, near England, and I am concerned to hear that the author of so much abuse of me here has sent to England by the same vessel in which he sailed, some virulently abusive pamphlet against me and him, to be published as soon as possible after his arrival." Cf. Cobbett, *Remarks on the Explanation lately published by Dr. Priestley, respecting the intercepted letters of his friend and disciple, John H. Stone. To which is added a certificate of civism for Joseph Priestley, Jun.* (London, 1799). This pamphlet was clearly intended further to blacken Priestley's reputation, and that of his son. It included two anecdotes of young Priestley's anti-English, anti-monarchical and pro-French sentiments, which he had apparently openly voiced when in Philadelphia in 1793, and which Cobbett published in the *Gazette* in January 1799, challenging Priestley to retract them: *Remarks*, 48–52. For Joseph Priestley, Junior's undoubtedly very extreme political opinions, cf. above, n. 190. For his departure for England, cf. J. Priestley, Jr. to J. Watt, Jr., 19 December 1798; T. Cooper to J. Watt, Jr., 4 January 1799, B. R. L.; and Priestley to Wilkinson, 25 December 1798, W. P. L. For his arrival, cf. J. Priestley, Jr. to J. Watt, Jr., 5 May 1799, B. R. L.; Priestley to Lindsey, 6 June 1799, *Works*, I.2.419; and Priestley to Wilkinson, 14 June 1799, W. P. L. For his return to America, in the late summer of 1800, cf. J. Priestley, Jr. to J. Watt, Jr. (May 1800), B. R. L.; and below, n. 392.

correspondence when apparently taxed upon his political stance: "Pray what have I written on *Politicks* since I came into this country," he wrote to John Vaughan, shortly before his son set sail, "besides a short letter about the *intercepted letters;* and yet I see them now advertized as printed in Philadelphia. As to Cobbet" (sic), he added, "I am sorry that any of my friends should have bespoke his favour. His abuse was as agreeable to me."[303] He was conscious as never before of his unpopularity in Philadelphia, and wrote to the Rev. Toulmin in England describing it: "I am considered as a citizen of France, and the rage against every thing relating to France and French principles as they say, is not to be described. It is even more violent than with you. This is a change that I was far from expecting when I came hither." He described, however, the other "party in the country," to which, he believed, many of the farmers in Pennsylvania belonged, who opposed the present Administration. "In Kentucky, where your son is Secretary of State, they are almost universally of it." The opponents of the Administration were, he believed,

so much opposed to the measures of the general government, that I begin to fear a division of the country, and perhaps a civil war will be the consequence. In my opinion, an amicable separation will be desirable, as the southern states in general are disaffected. In this state of Pennsylvania, the majority, I believe, are so too, though those who are so say little. The true state of the case will appear at the next election of a Governor, which will be the next autumn.[304]

To Lindsey, Priestley wrote that "we expect a warm session of Congress," and reported the news of the Resolutions from Kentucky, rejecting "the late acts of Congress respecting sedition and aliens, and sent to all the other state-legislatures for their concurrence. Their declaration on the subject is," he approvingly added, "for forcible composition, equal to any thing I have ever read. We suppose it to be drawn up by our friend Mr. Toulmin, the secretary of state." He repeated his belief that "a great majority of the people in these parts, and, I believe, through this state, disapprove of the late measures," but that this would "only appear in new elections for members of Congress, &c."; and he reported, not for the first time, that "party-spirit runs very high, and individuals are much exasperated against each other," and that it was for this reason, among others, that he was not going to Philadelphia for the winter.[305]

---

[303] Priestley to J. Vaughan (December 1798), A. P. S., Priestley Papers, B. P. 931.

[304] Priestley to Rev. Dr. Toulmin, 9 January 1799, *Works*, I.2.412–14.

[305] Priestley to Lindsey, 23 December 1798, ibid., I.2.412. It was Jefferson who was the "principal author" of the Kentucky Resolutions, which were passed through the Kentucky legislature on 10 and 13 November 1798: A. Koch and H. Ammon, "The Virginia and Kentucky Resolutions," 147, 154–6; Peterson, *Jefferson*, 611–16; Ford, ed., *Works*, VIII. 2.458–79. Priestley in his letter to Rev. Toulmin in England seems to have been well informed – almost certainly by Harry Toulmin, with whom he was in correspondence – of the "numerous public meetings" held in Kentucky during the summer to protest against the Acts (Koch and Ammon, ibid.; and see also J. M. Smith, "The Grass Roots Origins of the Kentucky Resolutions," *W. M. Q.*, Series 3, 27.1 [Jan.–April 1970]: 221–45). His erroneous surmise as to Toulmin's authorship however suggests that it was not from him that he heard the news about the passing of the Resolutions.

Stimulated very clearly by evidence of the widespread and increasingly vocal opposition to the measures of the Administration, drawing his customary radical conclusions from the ferment of opinion, and anticipating with something like enthusiasm the prospect of elections in which the popular voice would make itself heard, Priestley continued to correspond with his Federalist friend in Philadelphia, George Thatcher. In December he allowed himself an outburst against France—"that abominable country."[306] On 7 January however he criticised the policies of the Federalists in America in perhaps the frankest manner yet, and with an extraordinary disregard for the consequences. "You say," he wrote,

you wish I were as zealous a friend of America as Mr. Hone is of France. Both Mr. Hone and myself, as well as Dr. Price and many others, were as zealous in the cause of America as he now is in that of France. If I had not been so, I should not have come hither, nor am I changed at all. I like the country and the constitution of your government as much as ever. The change, dear sir, is in you. It is clear to me that you have violated your constitution in several essential articles, and act upon maxims by which you may defeat the whole object of it. Mr. Adams openly disapproves the most fundamental article of it, viz., *the election of the Executive*. But as you say, we cannot see our own prejudices, and cherish them as truths.

I may be doing wrong in writing so freely, and I have been desired to be cautious with respect to what I write to *you*. But I am not used to secrecy or caution, and I cannot adopt a new system of conduct now. There is no person in this country to whom I write on the subject of Politics besides yourself, nor do I recollect what I have written; but I do not care who sees what I write or knows what I think on any subject. You may, if you please, show all my letters to Mr. Adams himself. I like his address on the opening of the Congress, and I much approve of his conduct in several respects. I like him better than your late President. He is more undisguised. We easily know what he thinks and what he would do, but I think his answers to several of the addresses are mere intemperate railing, unworthy of a statesman.

Priestley then proceeded to defend his general philosophy of government, and to warn America, with the moral weight of one who was an acknowledged authority—by, as he said, Adams himself—on the subject, of the misfortunes she was preparing for herself:

My general maxims of policy are, I believe, peculiar to myself. When I mentioned them to Mr. Adams, he was pleased to say that "if any nation could govern itself by them, it would command the world." Of this I am fully persuaded; but he has departed very far from them. All that I can expect is the fate of the poet Lee, who, when he was confined in a mad-house, and was asked by some stranger why he was sent thither, replied, "I said the world was mad, and the world said I was mad, and they outvoted me." My plan would prevent all war, and almost all taxes. But if the calamities of war, heavy taxation, the pestilence, &c., or any other evil, be required for the discipline of nations, as I believe that in the present state of things they are, they will be introduced from some cause or other. This country as well as others wants a scourge, and you are preparing one for yourselves.[307]

---

306 Priestley to Thatcher, 20 December 1798, *P. M. H. S.*, Series 2, Vol. 3 (June 1886): 24.
307 Priestley to Thatcher, 7 January 1799, ibid., 25-6.

By the beginning of 1799 Priestley, increasingly incensed, as were so many, by the illiberal measures of Adams's Administration, and personally outraged by the diatribes of Cobbett, was demonstrating once more that fearlessness of the consequences which Rutt described as characterising some at least of his conduct at an earlier date. Some of his increasing outspokenness can surely be attributed to the fact that yet another restraining influence – that of his son Joseph – had been temporarily removed, leaving Priestley much in the company of Thomas Cooper, who at this time was living as a member of his household, and who was himself, in the spring of 1799, "up to the ears in my old pursuit," as he later wrote to Watt.[308] Thomas Cooper's activities against the Adams Administration were to lead to his arrest, trial, and imprisonment for six months under the Sedition Act.

In his first years in America Cooper, with Priestley, had sedulously avoided being drawn into the American political scene. "I suppose like me you have done with Politics," he wrote from Philadelphia to his young friend in England: "I get as many Causes & as many fees as I can, I drink & laugh with all parties & am regarded as an honest kind of fellow, & in train to become a respectable Citizen." He could not even at this time resist a comment upon the politics of his adopted country, however. On the presidential election in 1797 he wrote of the "strong contest here for the Presidency," of Adams's slight majority, and the choice of Jefferson as Vice-President: "Adams," he wrote, "is certainly a superannuated old Aristocrat. But," he added, "it is of little consequence in my opinion who is chosen." He expressed concern for the fate of his friends in England – "we look every post for news that you are all either clapt into Jail by the Aristocrats of yr. own Country, or upon the point of having yr. throats cut by the Democrats of France." He believed nevertheless that France was in the right for threatening America "with retaliation for permitting the British to seize vessels bound to France . . . but the President elect," as he rightly forecast, "will have a different card to play."[309]

In 1797, Thomas Cooper could still write that "good Sense and good intentions are enough." In the following year, however, after his rebuff at the hands of John Adams for a government post, although he still affected an aloofness from the political scene, his prejudices and inclinations were becoming increasingly evident. His appointment at the bar in Pennsylvania had, he wrote, been opposed by "a systematic opposition" from "the violent *federal* party (i.e. the favourers of the British opposers of the french Interests & favourites at John Adams's Court)," and "they gave as a reason among themselves that if a new district was appointed that damned democrat Cooper wd. be appointed Judge." This

[308] T. Cooper to J. Watt, Jr., 1 February 1801, B. R. L.; Priestley, *Memoirs*, I.200–1.

[309] T. Cooper to J. Watt, Jr., 3 January 1797, B. R. L.; and cf. also same to same, 4 April 1796, ibid.: "I am grown very mild and moderate in my politics, and equally dislike the democrats and aristocrats."

loss of a potentially lucrative post can have done little to endear the Federalists to Cooper, and although he wrote to Watt that he was "too much tired of politics to dwell on them to you," yet he devoted some space to his own current thinking:

I detest . . . the ambitious intermeddling of the french & the Conduct of the French & English towards neutral nations particularly our own. The people here are extremely exasperated against the Conduct of France; & are very ready not to say desirous of going to war with them: President Adams & his party are ferociously vindictive against them; but he seems to me to lean towards Monarchy, & to be a very weak & not a very well meaning Man. I do not trim between the parties,

Cooper could still assert just a month before the passage of the Alien and Sedition Acts, "but I laugh at them."[310]

In April 1799, however, Thomas Cooper was prevailed upon to undertake for a short time the editorship of the *Northumberland Gazette*. From April 20 until June 29, "all the letters and miscellaneous articles" which appeared in this apparently innocuous local newspaper were, with two minor exceptions, composed by him,[311] and they constituted a violently hostile attack upon the measures of Adams's Administration. Among them were two papers on "Political Arithmetic" in which, to a greater extent even than Priestley in his *Maxims* of 1798, Cooper—the English emigrant so conscious of the claims and inherent value of commerce in his native country—now presented a carefully argued case against the "commercial system" for the developing economy of America. "It seems determined in America," wrote Cooper,

that we shall be a COMMERCIAL country. Our navy, our army, our loans, our increased taxes, have arisen from our commerce. This is cried up as our most important resource; as the means of riches, of power, of consideration. Upon this ground are our present warlike exertions triumphantly defended. I, on the contrary, am firmly persuaded, *until* the home territory of a country be accurately cultivated, and fully peopled—until manufactures, founded upon population, are in a state to require other markets to be sought—that foreign commerce is a losing concern; an appropriation of capital in all cases inexpedient, and in most cases detrimental to the country; that it has proved so to the commercial nations of Europe: that to afford it support by prohibitions and bounties, or protection by engaging in wars on account of it, or by manning navies in its defence, is egregious folly and gross injustice. That if it cannot protect itself, or be carried on without the fostering aid of government, it ought, like every other losing scheme, to be left to its own fate, without taxing the rest of the community and their posterity for its support. That foreign commerce is particularly inexpedient in this country, where there is so much land calling aloud for cultivation and for capital, and so deplorably managed for want of these. . . . If any profession is to be fostered,

310 T. Cooper to J. Watt, Jr., 3 January 1797, May 1798, ibid.
311 Malone, *Cooper*, 91–2; cf. also Smith, *Freedom's Fetters*, 307–9.

Cooper concluded, "let it be the tiller of the earth, the fountain head of all wealth, and all power, and all prosperity."[312]

Cooper's *Essays on Political Arithmetic* were to be included in an edition of several of his articles for the *Northumberland Gazette*, which he published as a pamphlet in July 1799, with a Preface in which he sadly denounced the tendency of the measures of Adams's Administration. "I hope they will afford some proof," he wrote,

> that I remain in this Country what I was in Europe, a decided opposer of political restrictions on the Liberty of the Press, and a sincere friend to those first principles of republican Government, the Sovereignty of the people and the responsibility of their servants. Having adopted these opinions on mature consideration, and the fullest conviction, I shall retain and profess them; but I am sorry to say they are likely ere long to become as unfashionable in this Country as in the Monarchies of Europe.[313]

Cooper's *Essays* culminated with an Address, published on 29 June 1799, written with all the eloquence, argumentative flair, and mastery of the issues involved which had characterised his career as a radical propagandist in England, and were now to be devoted to the politics of America. There was, he wrote, with an appearance of temperate impartiality at the outset, "a party in this country accused of an indiscriminate opposition to the measures of government; who, in their turn" accused their opponents of "an indiscriminate support of every measure calculated to increase the power of the Executive at the expence of the interest of the country." Such accusations were doubtless exaggerated on both sides; nevertheless, he wrote,

---

[312] T. Cooper, *Political Essays* (2nd edn., Philadelphia, 1800), 32–50: "Prohibit nothing, but protect no speculation, no investment of capital at an expence beyond its national value. If wars are necessarily attendant upon commerce, it is far wiser to dispense with it; to imitate the Chinese and other nations who have flourished without foreign trade." (Ibid.) Malone (99, n. 70) points out that the first (and much the shorter) of these articles is signed "Back Country Farmer"; the second is initialled T. C. He concludes however that Cooper almost certainly wrote them both. McCoy, *The Elusive Republic*, 177, writes that Cooper's sentiments expressed in *Political Arithmetic* were "echoed in the late 1790s by scores of Jeffersonians, including Joseph Priestley, who wrote in the *National Magazine* under the name of 'A Back-Country Farmer.'" It was, however, Priestley in the *Maxims* (above, n. 278), which McCoy does not cite, who had first articulated Cooper's approach. While denying influencing Cooper, Priestley later fully endorsed the *Essays* and was, as he admitted, much in Cooper's company when he composed them: *Letters* (1799), *Works*, XXV.174; and above, n. 308, below, n. 359. In the absence of any conclusive evidence (which McCoy does not give) it seems impossible to judge whether he was the anonymous author. J. Appleby, *Capitalism and a New Social Order. The Republican Vision of the 1790s* (New York, 1984), 88–9, 92–3, similarly omits to mention Priestley in describing Thomas Cooper's influential contribution to the Jeffersonian position. It is a measure perhaps of the difficulty of defining very exactly what, in the crucial years 1799–1800, this position actually was, that both McCoy and Appleby can find in Cooper's *Essays* an endorsement of their rather different interpretations of Jefferson's economics. However, as McCoy points out (175–8) the Jeffersonian analysis was very close to Adam Smith's basic contention in *The Wealth of Nations*, that "ideally . . . no capital should be invested in commerce or manufacturing until a country's agriculture is fully developed"; (cf. also above, n. 278). The contributions of both Priestley and Cooper were to be warmly welcomed by Jefferson in the campaign of 1800 (below, n. 368). Cf. also Banning, "Jeffersonian Ideology Revisited," 19, n. 46.

[313] T. Cooper, *Political Essays*, (1st edn., Philadelphia, 1799), Preface, 10 July 1799.

I cannot help thinking that of late years, measures have been adopted and opinions sanctioned in this country, which have an evident tendency to stretch to the utmost the constitutional authority of our Executive, and to introduce the political evils of those European governments whose principles we have rejected.

He did not wish, he disingenuously (and clearly with the Sedition Act in mind) remarked, to impugn the motives of those who were responsible: rather to illustrate the tendency of the measures which were at present being pursued. This, he asserted, was "a fair object of decent and temperate discussion." He could, he continued, best illustrate his meaning by supposing that he were himself "in the President's chair, at the head of a party in this country; aiming to extend the influence of the governing powers at the expence of the governed; to increase the authority and prerogative of the Executive, and to reduce by degrees to a mere name, the influence of the people. How," he asked, "should I set about it? What system should I pursue?" And he proceeded to enumerate all the measures of Adams's Administration, its tendency to demote the authority of the state governments; its curtailing the liberty of the press by multiplying the laws against libel and sedition; its consistent denigration of "the doctrine of the Rights of Man, and the Sovereignty of the People," and its open contempt for the government of France. "I would decidedly prefer," he continued,

the nations whose government inclined to despotism, and treat with coldness and reserve, republics founded on the same basis with our own. Every known friend to those principles I would carefully discountenance, and prohibit the emigration hither of every foreigner who might be suspected of attachment to them. They should be the constant theme of abuse in the prints which I should deem it prudent to encourage, and the companies which my partizans should frequent.

And, he continued, warming to his theme, but undoubtedly treading now upon dangerous ground, "the more completely to enlist the ambitious, the needy and the fashionable. . . . I would take care it should be known that no place, no job, no countenance might be expected by any but those whose opinions and language were implicitly and actively coincident with my own." He would patronize the conventional forms of religion, and their practitioners; and he would find it to be

my evident interest to cultivate the monied men of the country; hence I would shew a decided preference to mercantile people and to the mercantile interest over the agricultural . . . Hence too I would encourage the Banking and the Funding systems. The latter particularly, because the more money I could borrow on any pretence, the more jobs, the more contracts, the more means should I have, of gaining over adversaries and rewarding partizans.

And as a further extension of this system, and to enable a president very effectually to curtail the liberties of his people, he would introduce a standing army. "The grand engine, the most useful instrument of despotic ambition, would be a standing army." A standing army, wrote Cooper, renders a Militia idle, and therefore useless and contemptible. It provides for the partizans of government, it arms the partizans of government, it disarms, it paralyzes their opponents. Hence the predilection of the monarchies of Europe for standing armies; not to defend themselves against invaders from without, but against the friends and principles of liberty from within.

To this end, he added, alarms should be manufactured, and invasion threats spread. "With the same view," he continued,

I would encourage a naval armament. . . . By a navy I lay hold of the popular prejudices of the people; I can assist in many ways a monarchy hostile to liberty against republicans;—I gain over to a man the mercantile interest for whose protection it is ostensibly (and indeed in great measures really) raised: and it furnishes an opportunity of commanding the sea-port towns of any state, who might venture a more active opposition to my views than I could safely submit to.

Such, wrote Cooper, in this Address which was to be widely distributed and publicised, provide valuable ammunition for the democratic opponents of the Administration, and arouse a corresponding wrath among the Federalists, "appear to me the obvious measures for a man to adopt placed in a situation to aim at power independent of the people." He accused, he said, no one of doing so: the Administration's measures had been adopted "after fair discussion, and sanctioned by the highest constitutional authority of the people." Nevertheless, "we have adopted measures that, were I placed in the situation above described, I should sedulously have promoted." And he reserved for the last the vesting of the right of making treaties, not, as the Constitution provided for, in Congress, but with the Executive. "The doctrine of *Confidence* in the Executive has been urged in this country with almost as much perseverance as by the friends of Mr. Pitt in England," he wrote. All fair opposition to government measures had been treated as a sign of disaffection; the Alien Law was "calculated to operate against the emigration of persons hostile to the tyranny of Europe," and "enacted in evident opposition to the language and principles of the constitution." The Sedition Law was passed "directly in the teeth of what has usually been conceived the plain meaning of that Constitution." Its doctrine of *"implied powers . . .* permits any stretch of authority to be assumed and defended." "Nor," he added,

can the forcible (not to say violent) language of the President against the principles of freedom, adopted and propagated by the French nation in common with ourselves—doctrines, at the head of which are the Sovereignty of the People, and Rights of Man, be otherwise than grating to the true friends of our Constitution. However good his intentions, too much I think has been urged, in his answers to addresses, against French principles and false philosophy. Principles and philosophy, which (however abused) will stand the test of all the argument, all the sarcasm, and all the declamation of their opponents, whoever they may be.

All the measures of Adams's Administration, Cooper concluded—and, he said, he could enumerate many more—were coincident "with what a leader inclined to despotism might wish."[314]

---

[314] T. Cooper, *Political Essays* (2nd edn., Philadelphia, 1800), 24–32.

Thomas Cooper's Address of 29 June was republished in the *Aurora*,[315] now under the editorship of the Irishman, Duane, on 12 July. It was this publication, and also the fact that it was at the same time "printed in handbills, and distributed" by none other than Priestley himself, that brought both Englishmen to the attention of the leading men in the government, convincing the Secretary of State, Pickering, that Priestley's "discontented and turbulent spirit" would "never be quiet under the freest government on earth," and apparently giving ample proof of the conversation which Benjamin Vaughan was alleged to have had, on his arrival in America, with Abigail Adams. Vaughan had said, the President wrote in reply to Pickering's account of Priestley's and Cooper's political activities, "that Mr. Cooper was a rash man, and had led Dr. Priestley into all his errors in England, and he feared would lead him into others in America." Adams told Pickering also of Cooper's and Priestley's application to him in 1797: "the disappointed candidate is now, it seems, indulging his revenge. A meaner, a more artful, or a more malicious libel has not appeared. As far as it alludes to me, I despise it; but I have no doubt it is a libel against the whole government, and as such ought to be prosecuted."[316]

It was not for this attack upon the government that Thomas Cooper was to be successfully prosecuted under the Sedition Act, but for an even more inflammatory attack upon Adams which he made in reply to one of his opponents in the following November, as he joined in the heated contest for the governorship of Pennsylvania. But Benjamin Vaughan's fears of his influence – if, indeed, he had expressed them in precisely that form – were to be amply justified in the months that followed. For throughout the summer of 1799 both Cooper and Priestley demonstrated an extraordinary disregard for the consequences in their political actions, in an atmosphere which they were clearly aware was fraught with much danger for them.

On 3 May Priestley wrote in unusually frank terms to Lindsey of the interest which he took in the political scene in America. A great number of the people in his neighbourhood, he wrote, were possessed of "as good sense as I ever met with anywhere," and "a great majority of them think as I do on the subject of *Politics*, which is perhaps more attended to here than with you, and in consequence of it party spirit is more violent. Both the parties," he continued,

are mustering all their force against the election of a new governor of this state, which takes place the next autumn, as they will for a *President* the next year. I have no doubt, but that the democrats will have a great majority in this state, tho the more wealthy, and all connected with the government, are on the other side, and almost all the Newspapers are in their hands.[317]

---

[315] *Aurora*, 12 July 1799.

[316] Pickering to Adams, 1 August 1799, Adams, *Works*, IX.5–6; Adams to Pickering, 13 August 1799, ibid., IX.13–14. Smith, *Freedom's Fetters*, 309–12.

[317] Priestley to Lindsey, 3 May 1799, D. W. L. Mss., passage omitted in Rutt.

It was in this same letter, however, that Priestley wrote despondently of the failure of the enterprise of which he had hoped so much, and of the very probable reason for its demise: "The States," he wrote, "have refused to grant any thing to our College in this town. The Walls are raised, and so, I believe, it will remain. I suspect politics have influence here. I think to resign my presidentship of it."[318]

In the following month he wrote more pessimistically still, on hearing of the conviction of his old friend and publisher Joseph Johnson (sentenced to nine months' imprisonment and a fine of £50 for publishing Gilbert Wakefield's seditious remarks on the state of opinion in England). "I am glad to find by my son that Mr. Johnson comes off better than expected," wrote Priestley:

. . . He certainly did not deserve even this punishment. But we are following your example here as fast as we can, and I am more narrowly watched than ever I was in England, tho I take no part in their politics at all. But a bad name once acquired is not easily got rid of, and it is taken for granted that I must be a very fractious troublesome person, or I should never have been driven out of England. It is generally thought now, that France wishes to be on good terms with this country, and the generality of people wish it. But the leading people prefer a connexion with England, at all hazards. What turn things will take is quite uncertain. I am glad,

he concluded, "to be so far from the scene of Politics."[319]

Within days of writing thus, however—as Adams was shortly to be informed, and as Priestley himself later admitted—both he and Cooper attended the local "*democratic* assembly" on 4 July, at which, as Priestley described it, "republican or democratical toasts were drunk, and where the late measures of administration were not praised." Priestley, as he subsequently wrote, "approving of Mr. Cooper's Essays" in the *Gazette*, "contributed one dollar towards printing a few copies of one of them, before it was known they would all be reprinted in the form of a pamphlet." Although he asserted several times that he had had no part in the composition of Cooper's *Essays*, yet he admitted, as Pickering's informant also stated, that he did "carry a bundle" of this handbill, which was to be circulated through the town, "from the printer's to the house of a brother democrat in the town."[320] Soon after their appearance Priestley sent a copy of the *Essays* to Lindsey in England: "If they come to your hand, you will form some idea of the state of the country, and it is violently agitated at present, as much as England ever was; I think, more, as the two parties are more nearly balanced." In November he described for Lindsey "the very great contest in this state" for the governorship—in which Thomas Cooper played so combative and prominent a role—and the vic-

---

[318] Priestley to Lindsey, 3 May 1799, *Works*, I.2.419.
[319] Priestley to Lindsey, 25 June 1799, D. W. L. Mss., passage partly omitted in *Works*, I.2.420. For Wakefield, cf. *D. N. B.*; for Johnson, ibid.
[320] Priestley, *Letters to the Inhabitants of Northumberland*, *Works*, XXV.128-130; Pickering to Adams, 1 August 1799, Adams, *Works*, IX.5-6.

tory of the republicans: "Had it been otherwise, we should have been exposed to much insult. There is the true spirit of Church and King here, though under other names."[321] In December he informed Thatcher in Philadelphia of the contribution to this "violently agitated" political scene which he too, after much deliberation, had decided to make: "Mr. *Cooper's pamphlet*," wrote Priestley, "was sent, together with the copies of mine, to Mr. Campbell, bookseller in Market street, by a waggon which left this town yesterday, so that you may soon see them."[322]

The pamphlet of his own to which Priestley was now referring was the second part of his *Letters to the Inhabitants of Northumberland*, written, he frequently claimed, to counteract the unceasing abuse to which he was subjected, and the criticism to which his activities had led in many quarters. To some, indeed, the uncontradicted reports of Priestley's political activities had come as a severe shock: "I thought Dr. Priestley was so absorbed in philosophical pursuits as to have no space left in his mind for little *quotediannial* politics," wrote one of Adams's correspondents:

I absolutely thought the honor of our country, was in some degree concerned in protecting a great literary character, who had taken shelter among us from the cruel attacks of his gothic pursuers. You will say, perhaps it is not the first time that I have been deceived, by that mixed, Sphinx-like animal, a philosopher.[323]

Young Thomas Boylston Adams wrote in melancholy strain to his mother:

Cooper's address, valedictory, I now remember to have seen & read at the time it first appeared, and upon a second perusal I shall only say, that if Dr. Priestley could recommend such a man as Cooper to office, & assist in giving currency to such opinions as are here expressed, he deserves all that Porcupine ever wrote or anybody else could think against him. I had never heard of his meddling before in any of our political concerns. These exotic reputations,

added young Adams, "are slippiry things to build on. I find so little fame, that stands the test of all trials and all scrutiny, that I am sometimes disposed to become a cynic & carp indiscriminately at all that falls in my way."[324]

It was indeed the President himself who, throughout this period, resisted the many calls made to have his old friend deported under the Alien Act, and, according to Priestley's son's account, he enjoined Priestley to "abstain from saying anything on politics, lest he should get into difficulty."[325] In defiance of all such advice, however, Priestley in the

---

321 Priestley to Lindsey, 12/19 September 1799, D. W. L. Mss.: letter much altered with omissions in *Works*, I.2.421-2; and same to same, 14 November 1799, *Works*, I.2.423. And cf. Malone, *Cooper*, 101-3; J. H. Peeling, "Governor McKean and the Pennsylvania Jacobins, 1799-1808," *P. M. H. B.*, 54 (1930): 320-9.

322 Priestley to Thatcher, 12 December 1799, *P. M. H. S.*, Series 2, Vol. 3 (June 1886): 30.

323 B. Waterhouse to Adams, 15 August 1799, Mass. Hist. Soc., Adams Papers, Reel 396.

324 T. B. Adams to Abigail Adams, 16 September 1799, ibid., Reel 396; and cf. Smith, *Freedom's Fetters*, 312.

325 Adams to Pickering, 13 August 1799, Adams, *Works*, IX.14: "I do not think it wise to execute the alien law against poor Priestley at present. He is as weak as water, as unstable as Reuben, or the wind. His influence is not an atom in the world." If this judgment seems

autumn of 1799, proceeded with the publication of his *Letters to the Inhabitants of Northumberland*. "I have some thoughts of addressing a few *familiar letters* to the inhabitants of this town and neighbourhood," he wrote to Lindsey in September, in describing the harassment to which he was unceasingly subject. "The letters are written, but I hesitate about publishing them, and perhaps it may be better to lay under the odium, as it will hardly be attended with any personal inconvenience. Mr. Cooper advises me strongly to publish them, but I have no great reliance on his judgment."[326] It was in this same letter, however, that he wrote of having sent a copy of Cooper's *Essays* to Lindsey; and it seems very probable that (in spite of some further—and contradictory—evidence that Cooper was also enjoining moderation upon him at this time)[327] he was eventually persuaded by the arguments and example of one of whom he at this time undoubtedly saw much, and who was himself now increasingly immersed in the politics of America. There was, moreover, a further damaging diatribe from Cobbett to give Priestley any further impetus which might be needed to clear, as he hoped, his name and reputation.

On 26 August 1799 Cobbett gave advance warning of his further attack upon Priestley by re-publishing, in the *Gazette*, the whole of his original publication on the intercepted letters of the previous year.[328] On 28 August, he published, almost in full, the text of the pamphlet of which Priestley had heard when his son was leaving for England, which had been published in London in April.[329] Completed by Cobbett at the end of January 1799, as a reply to Priestley's own vindication of his role in the intercepted letters, and addressed to the people of Birmingham, it was a further attempt to damage Priestley's reputation, at the same time pouring much slander and invective upon the members of the Vaughan family. In it Cobbett repeated his "most unqualified contempt" for

---

unnecessarily harsh, it should be remembered that Adams had in 1797 received a note from Priestley (above, n. 264)—which he does, incidentally, seem to have treated with great confidence—but which nevertheless, in its revelation of what Priestley in public consistently denied—his urgent wish to leave America for France—did not show him in an entirely favourable light. For Adams's private attempts at this time to urge Priestley to keep silence, cf. Priestley, *Memoirs*, I.201-2. See also Smith, 173-4.

[326] Priestley to Lindsey, 12/19 September 1799, D. W. L. Mss., passage partly omitted in *Works*, I.2.421.

[327] Cf. below, n. 332.

[328] *Porcupine's Gazette*, 26 August 1799.

[329] Cf. above, n. 302 for Cobbett's *Remarks*; and Priestley to Lindsey, 12/19 September 1799, *Works*, I.2.421: "Mr. Cobbett the author of the malignant pamphlet on my emigration, and who has never omitted any opportunity of abusing me, has just published, in his newspaper, the whole of a pamphlet, which he says was published in London, in April last, on the intercepted letters." Cobbett's *Remarks* did indeed appear, although without the attack on Joseph Priestley, Junior, in the *Gazette* on 28 August 1799. The *Remarks* are printed in *Porcupine's Works* (IX.245-78), but with much editorial alteration. This includes the omission (as in the pages of the *Gazette*) of the attack on young Priestley. It also includes the changing of the date and place for Cobbett's "Introductory Address to the People of Birmingham," which in the *Remarks* and the *Gazette* is stated to be "Philadelphia, January 30, 1799" (in the *Works* it is altered to "Bustleton, 12th Sep. 1798"). An opening paragraph by Cobbett, explaining the reason for his delay in replying to Priestley's letter of 4 September, is also omitted.

Priestley, and his utter aversion to the "cool, placid, Priestleyan cant." He investigated, moreover, at some length and in damaging detail, the accusations brought against William Stone in England, his trial, the activities of his brother in France, and the part played in this potentially treasonable conspiracy by Benjamin Vaughan—now, as Cobbett triumphantly pointed out, unmasked by Priestley himself. He took occasion moreover to vilify not only Benjamin Vaughan—for his traitorous activities, his hypocritical acceptance of a seat in Parliament from his patron Lansdowne, in defiance of all his protestations against corruption—but also the patriarch of the family, Samuel. His activities in England, sneered Cobbett, merited only contempt; his stay in America made him similarly unwanted: "Like Priestley he was disappointed, neglected, and despised; and he at last left the country in a dudgeon, just," added Cobbett, "as Priestley will, the moment he can do it with a prospect of living elsewhere in safety and ease."[330]

"Here," wrote Priestley with stoicism of this latest mixture of inaccurate vitriol and damaging political allegation, from one whom he had by now not surprisingly come to see as his tormentor, "it cannot I think add much to the odium under which, chiefly by his means, I lay before." But he feared for its effect upon the Vaughans. "Benj.," he wrote, "strangely hopes to be concealed at Kennebeck and now his retreat will be fully known; but," he added, "I never saw any reason that he had to fear anything, or to leave England at all."[331] This disarming but damaging inability to recognise the dangers of which many of his friends were certainly acutely aware, was to be much in evidence in his correspondence with the Vaughan brothers—both now clearly alarmed by Priestley's increasing propensity for political controversy, and concerned to distance themselves from its effects.

In the early autumn of 1799, John Vaughan paid a visit to Priestley, and reported upon it to Benjamin in Maine. He wrote of Priestley's "regrets that you did not arrive first in Ama. as he would then have pitched his tent near you—at present removal is totally out of the question, & a journey to see you is contemplated as impossible." Thomas Cooper, added Vaughan, "by turning political writer of the *warmest complexion* on the Antif: Side has drawn upon himself considerable obloquy"; and Priestley, "by unguardedly but publicly patronizing the most obnoxious" of his publications, "by having it printed in handbills to distribute," had "subjected himself to remarks which do not tend, to render his Situation

---

[330] *Porcupine's Gazette*, 28 August 1799; and cf. also Cobbett, *Porcupine's Works*, IX.245–278.

[331] Priestley to Lindsey, 12/19 September 1799, D. W. L. Mss., passage omitted in *Works*. Priestley was clearly at this time aware of the further damage which Cobbett intended doing to his reputation in England. For the cartoon by Gillray, and the verse by Canning, *The New Morality*, which apparently inspired it, cf. Chaloner, 35; and M. D. George, *Catalogue of Political and Personal Satires*, VII (London, 1942), 468–472. "Priestley's a Saint, and Stone a Patriot Still," Canning's poem ran. Cobbett in his attack had also made much of Coleridge's valedictory ode (cf. below, n. 440).

FIGURE 17.   John Vaughan (1756–1841) by Thomas Sully. Courtesy of the American Philosophical Society.

FIGURE 18. Benjamin Vaughan (1751–1835) by Thomas Badger. Courtesy of Diana Vaughan Gibson.

the most pleasant." On Vaughan's hinting, however, his wish that Priestley "should remain quiet," Priestley had replied that, on the contrary, "he was resolved to publish, that he had prepar'd himself—Mr. C. & myself with difficulty prevailed upon him to Suspend the measure, & he will probably do nothing unless some Strong impression Seizes him. T. C.," continued Vaughan, "having taken his stand will I suppose go on," and in spite of his advice, Priestley, he believed, would almost certainly "be drawn in . . . & you know him well enough to be persuaded, that no idea of personal or family comfort will prevail over, what he may conceive to be his rights or his duty.—I have," John Vaughan concluded, "been made very uneasy." And he reported also that Priestley was "much Surprised" at Benjamin's uneasiness about his justification of him in his reply to Cobbett, "and thinks your discretion useless and totally contrary to his System."[332]

To this communication Benjamin Vaughan composed a justifiably heated reply: "Your account of our friend surprizes me notwithstanding what I have known of him," he wrote:

When he was with Lord L. I could only stop a publication for 6 months, though it was to hurt his patron with the public & the court, appear when it would.—I do not wonder that he cannot comprehend our motives on these delicate subjects, for I never could comprehend his. I cannot conceive, however,

he added,

that he will be meddled with here, except in newspapers: but if he will deal in print himself, he must bear to be attacked in print. Happy indeed would it have been for him, had he lived near us, instead of near his present fiery friends.—I shall write to him, but indirectly only attack his conduct, & convey to him my best advice under an easy cover.[333]

On 30 November John Vaughan reported: "Dr. Priestley has published." He reported to Benjamin (inaccurately) that Priestley had "quoted Cobbett's expression relative to you," and that he had written to him "requesting that in his Second number which is to contain his opinion of our Government, & of the acts of our administration he will not allow your name in any Shape to appear."[334] It was clearly in reply to some such remonstrance on his part that Priestley himself wrote to John Vaughan, justifying again his decision to publish, and professing himself mystified by his former pupil's conduct. "It was not your *brother* but your *father* that Porcupine alluded to as having *left this country and in a dudgeon,* and whose example he said I should follow," he wrote:

Since what you said to me about your brother when you were here, I have neither written to him, nor mentioned his name in writing to any other person; and as

---

[332] J. Vaughan to B. Vaughan, 8 October 1799, A. P. S., Vaughan Papers, B. V. 46 p.

[333] B. Vaughan to J. Vaughan, 27 October 1799, ibid., B. V. 46.

[334] J. Vaughan to B. Vaughan, 30 November 1799, ibid., B. V. 46 p; and cf. Priestley to Lindsey, 14 November 1799, *Works,* I.2.423: "I have just printed one pamphlet, and shall comprise all that I have to say in another, which is in the press."

you say so much depends upon his remaining *unknown,* I shall keep to the same conduct. But if he thought of remaining unknown in this country, why did he not enjoin secrecy upon me, when he wrote to me, as he did presently after his arrival. I had no suspicion of doing anything that could have offended him, or have interfered with any of his schemes. But tho what is passed cannot be recalled, I will engage that no more offence shall be given. With respect to *myself,* however,

he added,

you must allow me to follow my own judgement. Whatever you may think of Porcupine's abuse, it made a great impression to my prejudice in these parts, where his paper was taken by every federalist that could afford it. And there is not one person in a hundred that knows any thing of my writings, or my history. In consequence of his writings, I was actually considered as a very dangerous person by many who I had imagined would have known better. I, therefore, chose that they should know the worst of me from my own confession. I was sufficiently averse to meddle with the politics of this country, and without farther provocation, shall proceed no farther.[335]

It was in sending his brother a resumé of this letter, on 29 December, together with a copy of the second part of Priestley's *Letters,* that John Vaughan wrote sadly of his former mentor: "He cannot be reasoned with. Cooper," added Vaughan, clearly now, with his brother, convinced of the latter's influence upon Priestley,

has a game to play, to write himself into Consequence with a party—His violence creates Violence in the Neighborhood, the Dr. has got himself into the Vortex insensibly, & been (by those who want the Sanction of his name) urged to Measures, he had hitherto avoided.[336]

From Benjamin Vaughan, in a letter to Lansdowne, came a comment which well expressed the incredulity of many of Priestley's friends at actions which were in flagrant contradiction of all his previous declarations of intent of avoiding involvement in the politics of America.

Dr. P. has made a publication for which I much grieve. He constantly represented to me his tranquil life & his abstinence from politics, & has suddenly adopted the acts & sentiments of the most imprudent zealots in politics. He has no turn for discretion himself; indeed his system (?is) against it; & he is surrounded by persons who are crafty knaves or hot headed firebrands. I shall write to him, but with little hope of doing good to one so decided upon doing himself harm.[337]

---

[335] Priestley to J. Vaughan, 12 December 1799, A. P. S., Priestley Papers, B. P. 931.
[336] J. Vaughan to B. Vaughan, 29 December 1799, A. P. S. Vaughan Papers, B. V. 46 p.
[337] Sarah M. and B. Vaughan to Lansdowne, 25 January 1800, Bowood Mss.

PRIESTLEY'S *LETTERS TO THE INHABITANTS OF NORTHUMBERLAND* AND THE ELECTION OF JEFFERSON TO THE PRESIDENCY 1799–1801

The "Measures he had hitherto avoided," the publication of his *Letters to the Inhabitants of Northumberland*—drawn from him "by the gross abuse of our common persecutor," as he wrote on sending a copy of them to Benjamin Rush—did, Priestley frequently claimed, do his reputation much good in the neighbourhood where he now lay under such suspicion. Among a wider audience, however, the reception of them was mixed: "censured by many," as Priestley himself admitted in a letter to Wilkinson, although by others, apparently greatly admired.[338] Written as they were under the pressure of much harassment, sense of isolation, and fear of deportation from the country—the President, as Priestley wrote, having "been again and again called upon to carry into execution" the Alien Act against him;[339] and designed to reassure his neighbours that he was "not so very dangerous a person" as Cobbett and the Federalists alleged, the *Letters* nevertheless, in their determinedly frank admission of much that he had previously denied, alarmed and distressed many of his friends, and gave further ammunition to his enemies. The *Letters* were to put Priestley into the good graces of the more extreme spirits of the republican party—and that this was a part at least of his intention can be inferred from the fact that he sent copies of them to Jefferson. For others, however, the extraordinary admissions of effective duplicity in his dealings with the world became in themselves an issue, detracting from the professed aim of clearing his reputation, and for many robbing them of such merit as they did possess. As a result, the *Letters to the Inhabitants of Northumberland* have been among

---

[338] Priestley to Rush, 6 January 1800, Penn. Hist. Soc., Rush Mss., 63; Priestley to Wilkinson, 15 December 1800, W. P. L.; Priestley to Belsham, 15 May 1800, *Works*, I.2.434.

[339] Priestley, *Letters to the Inhabitants of Northumberland*, *Works*, XXV.114–15: "Let any person only view my house, my garden, my library, my laboratory," he wrote in one more than usually disingenuous passage, "and the other conveniences with which I am surrounded, and let him withal consider my age, and the little disposition that I have shewn to ramble any whither, and say whether any person . . . could remove with more difficulty, or with more loss, than I should do." (Ibid., 115.) If he was right to argue that deportation would be "cruel and unjust," Priestley's implicit denial of his frequent attempts to leave America for France was less than straightforward. Cf. ibid., 116–17, for his response to Cobbett's accusation that, "'like Mr. Vaughan, I shall leave this country in dudgeon the moment I can do it with a prospect of living elsewhere with safety and in ease.' You who know the provision I have made for spending my days in comfort here," he wrote, "are better judges of the probability of this than any person at a distance can be."

the least read, and certainly least discussed, of Priestley's political works. The valuable testimony which they provide of his political involvement, both in England and America, has been much overlooked; and the extent of his contribution to the Jeffersonian opposition to the Federalists still not sufficiently appreciated.[340]

The *Letters*, Priestley claimed in writing to Lindsey in November 1799, were largely intended (as had been his *Familiar Letters to the Inhabitants of Birmingham*) to mollify the antagonisms which he had aroused among his neighbours. But, he significantly added, they had given him "a good pretence for saying many things that I think may be useful here, if I can draw any attention to them. If not, I shall be satisfied in having done all that I have had any means or opportunity of doing." "I have at length published my thoughts on the politics of this country," he wrote to the Unitarian, James Freeman, in Boston, "with the same freedom with which I have been used to treat all other subjects."[341] In the *Letters* he did indeed not only admit without hesitation or qualification what he had publicly so strenuously denied—his lasting commitment to political involvement; proclaim without hesitation or inhibition his adherence to the cause of revolutionary France; and admit also to some active participation in American politics. He also, at some length, attempted to define for America what should be the true nature of her government; he was at pains to point out how under the Administration of Adams it had departed from this ideal; and he discussed, in some detail, how, in his view, it might be amended.

Much suspicion and dislike had fallen upon him in the prevailing political climate in America, Priestley wrote at the outset of the *Letters*, because he was known to be a citizen of France. This, however, he declared, in the circumstances in which it had been granted, he still considered "an honour to me, and think that I have more reason to be proud of it than of being a native of any country whatever. I wish," he added, "I had done more to deserve it." He was proud, he wrote, to have been associated with the Convention of 1792: many more of the Departments of France would, he asserted, have elected him to it had his disinclination not been made known. And he challenged the Americans to find fault with this:

When I came to this country, in the year 1794, I found the people in general in unison with me on this subject. On all public occasions, "Success to the arms of France" was never omitted among the toasts that were drunk . . . There was no complaint of *French* principles then, though they were the same that they are now. They were universally considered as the principles of general liberty, and the same with *American* principles, that is, republican, in opposition to monarchical.

---

[340] Cf., however, C. Bonwick, "Joseph Priestley: Emigrant and Jeffersonian," 15–17. For the omission of any mention of Priestley in Cunningham, *The Jeffersonian Republicans*, cf. below, n. 368.

[341] Priestley to Lindsey, 14 November 1799, *Works*, I.2.423; Priestley to Rev. James Freeman, 20 February 1800, *P. M. H. S.* Series 2, Vol. 3 (June 1886): 33–4.

The change, he wrote, was not "in me but in the people here; and considering that old men do not easily change their sentiments or attachments, if I must change, you must allow me more time."[342] There was, he believed, no essential difference between the governments of France and America. In both the commitment to democracy was essential; and although the atrocities of the rulers of France must be deplored, this was not a reason for condemning him, merely because he was a citizen of France. If, moreover, the principles of the government of France were to be reprobated, then those of America must be similarly censured. The critics of France, it must then be supposed, must wish to overturn the government of America. "This," he wrote, "would give me the greatest concern," and indeed give him cause for departure.[343]

From France, Priestley inferred, nothing was to be feared. His own political beliefs, he asserted, were equally innocuous; and he had, he maintained, a right to express them. Moreover, he declared, he had, from his great experience and constant attendance to such matters, every reason to be listened to.

Having never had much capacity for the more active pursuits of life, I had from very early years a turn for speculation on every subject that has come before me; and they have been very various, as my writings will shew. Among them, politics, in such a country as England, could not be excluded, any more than religion or philosophy. And being now old, and of course less active, I am more disposed to think, and, having more experience, I presume I am rather better qualified for it than ever.

And, he added: "At least I have, in the course of a very various life, had the means of acquiring some political knowledge." In Lansdowne's household, he wrote, "which was altogether a political house," he

daily saw, and conversed with, the first politicians, not only of England, but from all parts of Europe. And, independent of that connexion, I have had more or less intercourse with most of the political living characters whose names you have heard mentioned, and with many that you have not heard of.

He knew, he said, many of the leading men in France, both before and after the Revolution—such men as Turgot, Necker, Brissot, and the Duc de Rochefoucauld. Although his own writings had "related chiefly to theology, philosophy, or general literature, some of them are political"— and he cited specifically his *Essay on the First Principles of Government*, and his *Lectures on History and General Policy*. There were, he furthermore asserted, "few political publications of much note" that he had not read; and he had been personally acquainted with such men as the Abbé Raynal, and Adam Smith. "If, therefore, I have no knowledge of the subject of politics, it has not been for want of the means, or the opportunity of acquiring it." He would surely not, he further, and in defiance of all previous protestations, added,

---

[342] Priestley, *Letters, Works*, XXV.119.
[343] Ibid., 119–22.

have been thought of as a proper person for a member of the conventional assembly of France, chosen in what are reckoned the best times of their revolution, for the express purpose of forming a new constitution of government for that country, if I had not had some character for knowledge of this kind. My knowledge of theology, or chemistry, would not have recommended me to that situation.[344]

He cited his friendship with Washington and his close relationship with Franklin—"his letters to me would have made a very large volume";[345] and he reiterated that it was above all his very great experience in political matters—"since it is chiefly passion that misleads men's judgment with respect to it," and the passions of old men "are generally more under the command of reason than those of young men"—which made it the less presumptuous in him "now to think and write" upon them.[346]

On the subject of his political activities in America, Priestley admitted that there had indeed been some involvement on his part. He had attended the two dinners on 4 July—"one of them two years ago, in a grove near this town, and this year in another near Sunbury"; he had written the articles on *Political Arithmetic* for the *Aurora*, which he considered of sufficient importance to publish as an Appendix to the *Letters*; and, although he "never saw any of the papers" that Cooper wrote "till after their publication," he had approved of them. He had "contributed one dollar towards printing a few extra copies of one of them, before it was known they would all be reprinted in the form of a pamphlet"; and he had helped "to carry a bundle of them from the printer's to the house of a brother democrat" in the town. He might, he wrote in an extraordinary passage, "if I could think that it would avail me any thing . . . perhaps plead that, if I have done mischief in some respects, I have done good in others." But in the present climate of opinion, no mercy would, he realised, be shown to him on this account. His only other political contribution had been his reply to Cobbett's "virulent" attack.[347]

Priestley then proceeded to discuss Hurford Stone's letter to him, and in sharp contrast—as he freely admitted—to his previous denial of responsibility for the sentiments of his friend, he now "freely" acknowledged that his letters "gave me great pleasure; and the like," he added, "I have received from others before and since that time, written by the same hand, and in the same spirit, though no two men think exactly alike, or would express themselves in exactly the same manner." Stone, he added, had earnestly wished, as he had also,

for a reformation of abuses in the English government, in order to prevent an entire revolution, which we did not think was wanted there. He now sees, or

---

[344] Ibid., 123–5.

[345] Cf. "Revolutionary Philosopher, Part I," 48, n. 27, for the destruction of Priestley's correspondence in 1791: the letters from Franklin would have been destroyed then.

[346] *Letters, Works*, XXV.122–6. Cf. above, n. 107, and "Revolutionary Philosopher, Part I," passim, for the extent to which Priestley attempted to deny what he did in his *Letters* of 1799, finally admit.

[347] *Letters, Works*, XXV.128–30.

thinks he sees, that no such reformation is to be expected, and therefore wishes a revolution to take place, thinking it to be absolutely necessary for the good of the people. I own,

wrote Priestley,

that I am now inclined to his opinion. I sincerely wish (if the genuine spirit of the original constitution cannot be revived, which would no doubt be the best for that country) for some more radical change than I have hitherto thought necessary, though I wish it may be effected peaceably, and without the interference of any foreign power.

"Shocked," he continued,

at the enormities which have been committed in France, and which no persons lament so much as the friends of liberty in every country, it has become fashionable with many to exclaim against all revolutions, indiscriminately, and all the principles that lead to them . . . But surely they who hold this language must either be avowed advocates of arbitrary power, or have forgotten the state of France before the last revolution.[348]

It is not difficult to see why the *Letters to the Inhabitants of Northumberland* so distressed John and Benjamin Vaughan, and so "much disturbed" and "alarmed" Lindsey.[349] It is still hard to understand, making every allowance for the circumstances in which Priestley found himself, how he could have failed to comprehend the effect which so many admissions of what he had previously denied, so many damaging assertions of complicity in events which had had such terrible consequences (and which were at the heart of the violent debate now dividing America) and so many provocative and superogatory assertions of political belief, would have on his situation in the hysteria prevailing in 1799. As a testimony of political extremism, however, of unswerving, naive, even ruthless idealism, and as evidence of a revolutionary mind prepared to accept if not condone much in order to obtain the desired end, Priestley's *Letters* are undoubtedly of value. And in this latter respect, his views on the American Constitution, which form the bulk of the second part of the work, are also of considerable interest, as Colin Bonwick has already noted.[350] For if some of the specific criticisms which Priestley made were subsequently to be retracted, they serve nevertheless to demonstrate his fundamental political ideas, as well as providing evidence for the interest which, in spite of all his protestations, he unquestionably took in the politics of America. It is important, however, to see these too in their full historical context: to realise the extent to which Priestley himself had suffered from the abuse of the executive power under Adams, and to comprehend the gravity of the situation into which this interpreta-

---

[348] Ibid., 130–2.
[349] Cf. Priestley to Lindsey, 1, 29 May 1800, *Works*, I.2.432; 434–5: "I am prepared to receive a very severe censure of my 'Letters to the Inhabitants of Northumberland.'" And cf. Priestley to Belsham, 15 May 1800, *Works*, I.2.433.
[350] C. Bonwick, "Joseph Priestley, Emigrant and Jeffersonian," 15–18.

tion of the Constitution of 1787 had put him. Insofar, moreover, as he was writing, as was so often the case, for the politics of the moment—having "a good pretence," as he had expressed it to Lindsey, "for saying many things that I think may be useful here"—it is important also to bear in mind the contribution which, as he was certainly aware, his comments upon the Constitution were likely to make to the current political furor. As an eloquent defence of the freedom of the press—an essential plank of the Jeffersonian platform in the campaign of 1799–1800—they remain, even by Thomas Cooper's *Essays,* unsurpassed.

The American Constitution was, Priestley wrote at the outset of his critique, "the best that has ever been devised by man, and reduced to practice in any age, or in any part of the world. It has every thing that is valuable in the English constitution (which was confessedly superior to any other in Europe,) without its defects. Without this persuasion," he added, "I should not have come among you." He proposed, however, several amendments, the most important of which was to alter the potential eligibility of the office of President for life. Such a possibility must, he believed, inevitably invite corruption. Even the present constitution of France, although lacking "union, and consequently . . . energy," was probably superior, because of its division of executive power, to that of America at present. The argument of the fickleness of a constantly changing executive he countered with that of the inherent moral cohesiveness of republican government, based as it was upon the popular voice: "The leaders of a government truly republican, like that of the United States, will, and must, take their measures from the wishes of the people, which are not so apt to change, because they flow from the general interest."[351] He wrote of his dislike of the controlling power of the Senate, and quoted a passage from his own *Lectures on History and General Policy* on the "manifest absurdity" of having "any more than one will in any state; because when any part of the government has an absolute negative on the proceedings of the rest, all public business may be at a stand." The inconvenience, he wrote, could only be remedied by taking from the Senate

their absolute negative on the proceedings of the House of Representatives . . . The more temperate deliberations of a few may be an useful check upon the impetuosity with which popular speakers often impel the proceedings of a greater number; but they are not so likely to entertain the real sentiments and conform to the wishes of the people at large, as the persons who are their more immediate and their later choice.[352]

Priestley's sympathy with the Jeffersonian stance on states' rights against the maladministration of the general government, and with the prevailing Republican distrust of the Federalist judiciary, he made very clear in his advocacy of "a special court, consisting of deputies from all the

---

[351] *Letters, Works,* XXV.156–7.
[352] Ibid., 158–9. Cf. *Works,* XXIV.230; and above, n. 135 for his further elaboration of this point in his *Political Dialogue.*

states of the union," to guard against violations of the constitution "by persons entrusted with its administration." "The greatest danger of any encroachment on the constitution," he wrote, "is from the congress mistaking or exceeding their power." The judiciary, as "in all countries, and in all times," were likely to favour the existing administration. And he proposed that "it should be in the power of the legislatures of any of the separate states to call this special court, and lay before it whatever they may apprehend to have been a violation of the constitution, by the congress, the president, or any man, or body of men, whatever."[353] He opposed the practice of oath-taking for office. And in a further attack upon the encroachments of the executive power under Adams, he deplored the removal of the right of making treaties from the Representatives to the Senate and the President: it was, he asserted, an infringement of the constitution, and a dangerous one. So too, the denial of the Representatives' right to object to the raising of money to carry a treaty into effect, implied a violation of constitutional right which "even the parliament of Great Britain has not yet been brought to surrender. . . . It . . . appears most clearly to me," he wrote, "who am a stranger among you, that the real meaning and intent of the constitution in these two essential articles has been perverted."[354] And on this theme he attacked also with great vehemence the Alien and Sedition Acts:

Laws calculated to restrain the freedom of speech and of the press, which have always been made on the pretence of the abuse of them, are of so suspicious a nature in themselves, and have been so constantly the resort of arbitrary governments, that I was beyond measure astonished to find them introduced here; and yet in some respects the laws that have lately been made by Congress are more severe than those in England.

All attempts to muzzle a free press, Priestley wrote, were doomed to failure. In France under the old monarchy, he had himself, in 1774, been astonished at the extent of the censorship prevailing: "But did this strictness prevent the revolution? The freest publications," he wrote,

were at the same time circulated with the greatest industry, and they were read with avidity, and with tenfold effect, in consequence of it. The same will be the case in every other country in which the same measures shall be adopted; so that, without pretending to any extraordinary means of prying into futurity, we may predict that the cause of *monarchy* in England, and that of *federalism* in this country, will be no gainers eventually by what their advocates are doing in this way.[355]

The Alien Laws, he asserted, were expressly designed to keep out of America "the friends of liberty (opprobriously called *Jacobins*, *Democrats*, &c.) emigrating from Europe, a description of men in which I am proud to rank myself . . . Had those laws been made six years ago, there would not have been an Englishman in this place."[356] "To find in America," wrote Priestley,

---

[353] *Works*, XXV.159.
[354] Ibid., 160–3.
[355] Ibid., 163–5.
[356] Ibid., 165.

the same maxims of government, and the same proceedings, from which many of us fled from Europe, and to be reproached as disturbers of government there, and chiefly because we did what the court of England will never forgive in favour of liberty here, is, we own, a great disappointment to us, especially as we cannot now return. Had Dr. Price himself, the great friend of American liberty in England, or Dr. Wren,[357] with both of whom I zealously acted in behalf of your prisoners, who must otherwise have starved, and in every other way in which we could safely serve your cause, because we thought it the cause of liberty and justice, against tyranny and oppression; I say, had either of these zealous, and active, and certainly disinterested, friends of America been now living, they would not have been more welcome here than myself; and they would have held up their hands with astonishment to see many of the old Tories, the avowed enemies of your revolution, in greater favour than themselves.[358]

It was on this high moral ground—as one who had stood by America in her hour of need, and well understood the essential principles on which her Constitution had been founded; who claimed, moreover, that his political standing was such that he was "perhaps . . . in a great measure the cause of the prevailing jealousy of foreigners"—that Priestley proceeded, in his concluding letters, to reprobate further the tone and tendency of the Federalists' measures against France; to deplore their building up of a strong naval strength, and resort to a standing army, to the detriment of the true interests of the country. And he recommended on the important subject of *Political Arithmetic* the *Essays* of Thomas Cooper, "who," Priestley still claimed, "independently of me, adopted the same principles, and has enforced them in his excellent manner."[359]

If his English friends were dismayed by the publication of the *Letters*, reaction in America was mixed. They had, Priestley wrote in January to Lindsey, "completely answered my purpose *here*, though they have exposed me to more abuse at a distance; but that," he added, "gives me little concern."[360] To Russell, a few weeks later, he described however, the considerable impact which they had in some quarters in America, and referred for the first time to his communications with one of the most zealous of the opponents of Adams—Thomas Jefferson—on the subject. "What we printed here," he wrote,

Mr. Jefferson informs me, were all sold in a day or two, and Chancellor Livingston, of New York, says, he has printed an edition at Albany. I think it probable that a new edition will be printed at Philadelphia; so that you will easily get them. I hear there is an answer at New York; but I have not yet seen it. I shall hardly be tempted to do any thing more of this kind,

---

[357] Cf. ibid., 167, note, and McLachlan, *Letters of Lindsey*, 87, for Dr. Wren, Dissenting Minister of High Street Chapel, Portsmouth (1757-1787), who helped American prisoners-of-war in that town and was "instrumental in saving the lives of many hundreds as well as relieving thousands." Cf. also Sheldon S. Cohen, "Thomas Wren: Ministering Angel of Forton Prison," *P. M. H. B.*, 103 (1979): 279-301; C. M. Prelinger, "Benjamin Franklin and the American Prisoners of War in England during the American Revolution," *W. M. Q.*, 3rd Series, 32 (1957): 261-94, where, however, no mention is made of Priestley's participation.

[358] Priestley, *Letters, Works*, XXV.167.

[359] Ibid., 168-74.

[360] Priestley to Lindsey, 16 January 1800, *Works*, I.2.425.

he added, "but I do not repent of what I have done."[361] To Livingston, writing on the publication of his *Letters* in Albany, he wrote that they had, he found, "given the greatest offence to the Federalists, and to many, I hear, who are of the other party. . . . Having said what I thought necessary," he added, "I shall hardly repeat the offence. I am sure I wrote without anger, or ill will to anybody, and with a sincere respect for the constitution of the country."[362] In a subsequent letter to Russell he expressed his gratitude for his friend's understanding of what he had done: from another correspondent, he had had a letter "remonstrating in such a manner against my Letters, as if I had committed the sin against the Holy Ghost."[363] To Lindsey he wrote again that "at a distance, the abuse of me is not diminished," and he admitted further to Belsham that "some Federalists have said, that the other party, who wish me well, disapprove of my writing." And yet, he was able with satisfaction to point out, "others, and some of the first men in the country, say that my Letters have done much good, and have made a great impression on many, and they write to me in the most flattering terms on the subject." "I believe that on the whole, the Letters have done real good," he wrote again to Lindsey: "Being shut out from every thing in the way of theology, I see no reason why I should not endeavour to be useful in any other. But I believe I have done with the subject; and if I have offended, shall do so no more. As to —," he added, dismissing adverse reaction which had apparently reached Lindsey from other quarters: "You should not pay any attention to what he or the family says. They consort only with Federalists."[364]

Among those whose approval Priestley was gratified and surprised

---

[361] Priestley to Russell, 7 February 1800, *Works*, I.2.427; cf. Jefferson to Priestley, 18 January 1800 (below, n. 368): "The stock of them which Campbell had was, I believe, exhausted the first or second day of advertising them." For the "answer at New York," cf. below, n. 376.

[362] Priestley to Livingston, 17 April 1800, Schofield, *Scientific Autobiography*, 303. For Robert Livingston (1746–1813), "chancellor of New York, statesman, diplomat, farmer, experimenter," and correspondent of Priestley, cf. *D. A. B.*, Schofield, 364–5, and G. Dangerfield, *Chancellor Robert R. Livingston of New York (1746–1813)* (New York, 1960), 287, 295.

[363] Priestley to Russell, 3 April 1800, *Works*, I.2.430. It is impossible to know who this correspondent was, although it may have been one of the Vaughan brothers. Cf. J. Vaughan to B. Vaughan, 11 December 1799, A. P. S., B. V. 46 p: "Dr. P. second book. I have seen in part. When published I shall send both. In this, he gives his opinion of our administration—of course not favorable. He proposes amendmts to the Constitution

1st President for 3 or 4 ys. & Eligible again.

2. A Court, composed of members from each State to decide whether Laws are Constitutional or not & this Court to be Called together *by any one State.*

3. To Cease trading particularly in war time, to be like the Chinese."

On 1 December 1800, William Russell wrote, in the course of a long letter to Benjamin Vaughan, that, "in estimating the propriety of Dr. P'-s publication, I have no reference, either to him, or his Book, as holding him up as a *Political* Character, nor do I Conceive him, to have the least disposition, or wish to be considered as such—in Theology I desire to find him": and there, William Russell asserted, Priestley's contributions, albeit controversial, and such that Benjamin Vaughan might wish him to avoid, had been the means of enlightenment. (A. P. S., B. V. 46 p, William Russell to Benjamin Vaughan, 1 December 1800). William Russell's notion of "a political character" was, it must be said, as ambivalent as was Priestley's.

[364] Priestley to Lindsey, 1 May 1800, *Works*, I.2.432; Priestley to Belsham, 15 May 1800, ibid., I.2.434; and also Priestley to Lindsey, 29 May 1800, ibid., I.2.434–6.

to secure, was his Federalist correspondent in Philadelphia, George Thatcher. "Your letter was peculiarly welcome," Priestley wrote on 12 December 1799,

for in truth, I was afraid you would have revolted at my *Politics*; as you are so violent a Federalist, and I such a democrat. Since, however, you could bear the *first* part of my letters, I will venture to send you the *second* by this post, and then you will know the worst of me.

"If you dare trust me with any political information," he characteristically added, "I shall be glad to receive it."[365] On 9 January, he wrote again, on sending the second part of the *Letters*: "With respect to *this* I only ask your forbearance, and if you think I have offended, your forgiveness." He defended again his reason for writing them, declaring that they had been "of great use": but that, also, "I find there are many Porcupines in this country, and with them my case is not at all mended."[366] On 28 January, however, he could express his relief at Thatcher's reaction: "I truly admire the candour you express with respect to my *Letters*. It is more than I expected even from you, and I must say that with the Federalists in general, it is very uncommon. . . . I shall thank you," he added, "if you will tell me what you wished me to have *omitted* in the Letters, or anything in which you think I am *palpably* wrong. I do not mean to draw you into any controversy, but I will *think* of it."[367]

Clearly, however, the approbation that gave Priestley the greatest pleasure was that of Jefferson, the Vice-President whose open dissent from the measures of the Adams Administration had become a crusade to restore the Constitution which the Federalists were apparently bent upon undermining. Shortly after their publication, Priestley sent Jefferson copies of his *Letters* and Thomas Cooper's *Essays*, and from Philadelphia in January Jefferson wrote an appreciative reply:

You will know what I thought of them by my having before sent a dozen sets to Virginia to distribute among my friends. Yet I thank you not the less for these, which I value the more as they came from yourself. . . . The Papers of political arithmetic, both in your's and Mr. Cooper's pamphlets are the most precious gifts that can be made to us; for we are running navigation-mad, & commerce-mad, & navy-mad, which is worst of all. How desirable is it that you could pursue that subject for us. From the Porcupines of our country you will receive no thanks; but the great mass of our nation will edify & thank you.[368]

---

[365] Priestley to Thatcher, 12 December 1799, P. M. H. S., Series 2, Vol. 3 (June 1886): 29–30.

[366] Priestley to Thatcher, 9 January 1800, ibid., 31; and cf. also his letter of 23 January, below, n. 367: "As to the Federalists at a distance, I stand, as I expected, just as I did before."

[367] Priestley to Thatcher, 23 January 1800, ibid., 31–2. And cf. Priestley to Belsham, 15 May 1800, *Works*, I.2.433.

[368] Jefferson to Priestley, 18 January 1800, Lib. Cong. Mss., Jefferson Papers, and Ford, ed., *Works of Jefferson*, IX.95–9; and cf. also Priestley, *Works*, I.2.435–6. Cf. also Jefferson to Nicholas, 7 April 1800, Ford, ed., *Works*, IX.127–9, for Jefferson's further distribution, in April 1800, of eight dozen copies of Thomas Cooper's two "Essays on Political Arithmetic" in Virginia: "He was printing a 2d edition of the whole," Jefferson wrote, "& was prevailed on to strike off an extra number of the two on Political arithmetic, adding to it some principles of government from a former work of his." (This was Cooper's *Propositions Respecting*

Jefferson took this opportunity to commiserate with Priestley on the persecution to which he had been subjected: "How deeply have I been chagrined & mortified at the persecutions which fanaticism & monarchy have excited against you, even here!" It continued, he observed, even on the demise of Porcupine: "You have sinned against church & king, & can therefore never be forgiven." And he made abundantly clear the regard in which, by contrast, he held Priestley, by asking for his advice on the proposed University of Virginia: "The first step is to obtain a good plan," wrote Jefferson:

that is a judicious selection of the sciences, & a practicable grouping of some of them together, & ramifying of others, so as to adapt the professorships to our uses, & our means. In an institution meant chiefly for use, some branches of science, formerly esteemed, may be now omitted; so may others now valued in Europe, but useless to us for ages to come. . . . Now there is no one to whom this subject is so familiar as yourself. There is no one in the world who equally with yourself unites this full possession of the subject with such a knowledge of the state of our existence, as enables you to fit the garment to him who is to *pay* for it & to *wear* it. To you therefore we address our sollicitations.

Priestley's suggestions, Jefferson added, would be received by those who had already given thought to the subject—some of "the ablest and highest characters of our state"—"with the greatest deference & thankfulness." And he concluded his letter by urging Priestley to visit Virginia to meet Du Pont, the French *philosophe* and physiocrat whom in 1798 Adams had refused entry to America, and who was to exercise a great influence over Jefferson's thought: "I have no doubt the alarmists are already whetting their shafts for him also," Jefferson wrote, "but their glass is nearly run out; and the day I believe is approaching when we shall be as free to pursue what is true wisdom as the effects of their follies will permit; for some of them we shall be forced to wade through because we are immersed in them."[369]

In a second letter from Philadelphia Jefferson again urged Priestley to bear a little longer the harassment to which he had been subjected: "Pardon, I pray you, the temporary delirium which has been excited here, but which is fast passing away. The Gothic idea that we are to look backwards instead of forwards for the improvement of the human mind" was not, he believed, one that "this country will endure; and the moment of their showing it is fast ripening; and the signs of it will be their respect for you, & growing detestation of those who have dis-

---

the Foundation of Civil Government: above, n. 91.) Cf. Cunningham, *The Jeffersonian Republicans*, 221-2, where, however, no mention is made of Jefferson's appreciation and distribution of Priestley's pamphlet.

[369] Jefferson to Priestley, 18 January 1800, Lib. Cong. Mss., Jefferson Papers, and Ford, ed., *Works of Jefferson*, IX.95-9, where "immersed" is transcribed "emerged." For Priestley's "Hints Concerning Public Education," which he sent in response to Jefferson's request, cf. Priestley to Jefferson, 8 May 1800, Jefferson Papers; G. Chinard, *The Correspondence of Jefferson and Du Pont de Nemours* (Johns Hopkins Univ. Press, 1931), 16-18. And cf. Malone, *Jefferson and the Ordeal of Liberty*, 447-51, for Jefferson's anxious courting of Priestley at this time.

honored our country by endeavors to disturb your tranquility in it." And
he again wrote of his wish for a visit from both Du Pont—due to arrive
in Philadelphia very shortly—and Priestley, to Virginia, "to shew us two
such illustrious foreigners embracing each other in my country as the
asylum for whatever is great & good."[370]

Priestley transcribed for Lindsey Jefferson's letter of 18 January—"to
shew you that *all* persons here do not see them in the same light with
you." Lindsey, he added, would "likewise from this form some idea of
Mr. Jefferson, and his political principles, which is of some consequence,
as he is generally thought to be in many respects the first man in this
country, and will probably be our next President."[371] To Jefferson, however, he replied in tones of some despondency: "Indeed, it seems extraordinary, that in such a country as this, where there is no court to dazzle
men's eyes, maxims as plain as that 2 and 2 make 4 should not be understood, and acted upon. It is evident that the bulk of mankind are governed by something very different from reasoning and argument." Many,
he wrote in a subsequent letter, speak

of the increasing spread of republican principles in this country. I wish I could
see the effects of it. But I fear we flatter ourselves, and if I be rightly informed,
my poor *Letters* have done more harm than good. I can only say that I am a sincere
well wisher to the country, and the purity and stability of its constitution.[372]

To this Jefferson wrote a reassuring reply:

The mind of this country is daily settling at the point from which it has been led
astray during the (last few) years. I believe it will become what the friends of
equal rights had ever hoped it would: and I trust the day is not distant when
America will be proud of your presence, & be anxious only to find occasions of
obliterating the pain which some of her dogmaticians stimulated by and counte-
nanced by foreign malice, have been able to excite in your mind.[373]

Encouraged undoubtedly by these testimonies of his value to the
Republican party, Priestley was to arrange a second edition of the *Letters
to the Inhabitants of Northumberland* after Jefferson's election. He had been
told, he wrote in the Preface, "that, as far as circulation extended," they
had "contributed something" to the successful campaign of the Jeffersonians against Adams.[374] But if the *Letters* had given Priestley the stamp
of official republican approval, they had also aroused strong passions,
and confirmed old prejudices. "This eldest Son of Disorder will never
obtain his sought for 'Reform' on this Side the Grave," wrote one of his
opponents; "and I believe, the Government of Heaven itself, should he
ever get there, will, in his Opinion, want Reformation." In Philadelphia,

[370] Jefferson to Priestley, 27 January 1800, Lib. Cong. Mss., Jefferson Papers, and Ford,
ed., *Works of Jefferson*, IX.102-5, where "your tranquility" is transcribed "our."
[371] Priestley to Lindsey, 29 May 1800, *Works*, I.2.434-6.
[372] Priestley to Jefferson, 30 January, 8 May 1800, Lib. Cong. Mss., Jefferson Papers.
[373] Jefferson to Priestley, 11 August 1800, ibid.
[374] Priestley, *Letters, Works*, XXV. Preface, 111-12; and cf. Priestley to Belsham, 25 June
1801, *Works*, I.2.466; Priestley to Jefferson, 10 April 1801, Jefferson Papers, and Appendix.

Mathew Carey apparently wondered whether Priestley's outspokenness in politics had not affected the sale of his books.[375] And in one of the more thoughtful published attacks upon Priestley, Noah Webster reproached him for taking the liberty, as had Cobbett, of criticising the politics of a country in which he was a foreigner.

In his *Ten Letters to Dr. Priestley*, Webster attacked him without mercy for his ambivalent and disingenuous attitude towards France. Priestley, he wrote, had ignored the *"real* reason" for the change of sentiment in America, and wrote instead merely about a resemblance in the abstract principles of the American and French governments. It was the French themselves, wrote Webster, who had "abandoned the principles of their own Constitution"—as the people of America had not been slow to perceive—in their violation of the rights of their own people, and their arrogant aggression throughout Europe, an aggression which they patently hoped to pursue in America. The political experience of which Priestley boasted, Webster mocked: it had hardly, he wrote, led him into a sound system of politics. He took issue with Priestley on his critique of the Constitution: the Presidency, he asserted, was an office of such importance that it required "exalted talents." The time, labour and experience required to qualify a man for such an office would be wasted if he could not hold it for life; nor had experience shown that this was corrupting. The Special Court proposed by Priestley to bring the executive officers of government to account would, on the other hand, be open to abuse. And he claimed that Priestley had misrepresented the arrogation by the Executive of the treaty-making powers of the House of Representatives. The withdrawal of all support for the merchants of America, and the abolition of all marine defence, he declared utterly impracticable. Priestley's suggestion that Jay's Treaty should not have been made with England without the prior knowledge of the French, he described as degrading to the independent spirit of America. Of all the proposals made by Priestley for change, however, Webster singled out for the severest criticism the bland acceptance by the latter, in his *Maxims of Political Arithmetic*, of the possibility of a civil war. This must, wrote Webster, be the boldest plan for a political opposition ever seen. Priestley, he counselled, should in political matters remain silent.[376]

Throughout 1800, as Jeffersonians and Federalists joined battle for the forthcoming presidential election, Priestley was the target of much further and bitter abuse. To Webster, after some hesitation, he composed

---

[375] Nicholas Ridgely to ?, 21 December 1799, Penn. Hist. Soc., Gratz Colln., Case 3, Box 32; Priestley to Carey, 25 June 1801, Penn. Hist. Soc., Lea and Febiger Colln.: "I hope that the reason you suggest for the little demand for the *Comparison* is not just, as it will imply that there are no purchasers of books besides Federalists; and I hope that there are even among (*them*: illegible) some who would not think the worse of me as a philosopher, or a theologian, merely on account of my differing from them in Politics."

[376] N. Webster, Jun., *Ten Letters to Dr. Priestley, in Answer to his Letters to the Inhabitants of Northumberland* (New Haven, 1800). For Webster, a former sympathiser with revolutionary France, cf. Tagg, *Aurora*, 145-6.

a private reply, which he sent to Thatcher for approval. "I wish . . . you would read the letter, and if you think there is anything in it, that you would not have me to say to him, suppress it."[377] And this new found caution, in spite of an assurance from Thatcher that "the *reign of terror*," as Priestley described it, was drawing to an end, can be seen in his correspondence with the Vaughans. He had, he wrote to John Vaughan in July, had two letters from Benjamin, "which require an answer, tho I had thought it would be better (for the reasons I gave you) to suspend our correspondence till a change of time and circumstance should make it less hazardous."[378] In April 1800 there had come a further farrago of denunciation from Cobbett, mocking Priestley now above all for the errors which he had found in the Constitution of America, and urging him to admit his mistake, and return to England. In the *Gazette of the United States*, an extract of a letter from John Quincy Adams was published attacking the "Jacobins" Priestley and Cooper.[379]

With the defeat of the Federalists in the Election in the autumn of 1800, much of the abuse of Priestley did, as Jefferson had predicted, die away. To his friends in England, Priestley expressed his great sense of relief: "The election of Mr. Jefferson for our next President is now secure by a considerable majority," he wrote to Lindsey, albeit somewhat prematurely, in December,

and also that of Mr. Burr, of New York, with whom I am well acquainted, for Vice-president. The measures of the late administration are now almost universally reprobated, and I hope, in due time, some strong censures, such as will prevent their future occurrence, will be passed upon them. But Mr. Jefferson will do nothing rashly. His being President,

he added, "may induce me to visit the federal city, and perhaps his seat in Virginia."[380] Pressed frequently by Jefferson in the next few years to visit him at Monticello, Priestley, largely on the grounds of ill health—for a severe fever afflicted him in Philadelphia in the spring of 1801—never undertook the arduous journey to the more temperate climes of Virginia. That he increasingly realised, however, the danger in which he had stood under the Administration of Adams is clear from one of the letters which he wrote from Philadelphia in March 1801: "You will rejoice with the friends of liberty in this country," he wrote to Belsham, after the potentially very damaging deadlock between Jefferson and Burr had been finally resolved,

---

[377] Priestley to Thatcher, 6, 20 March, 23 April 1800, *P. M. H. S.*, Series 2, Vol. 3 (June 1886): 34, 36.

[378] Priestley to Thatcher, 10 May 1800, ibid., 37–8; Priestley to J. Vaughan, 7 July 1800, A. P. S., Priestley Papers, B. P. 931.

[379] Cobbett, *Porcupine's Works*, XI.395–434: 30 April 1800: Letter from P. Porcupine to Dr. Priestley; Smith, *Freedom's Fetters*, 312, n. 8.

[380] Priestley to Lindsey, 25 December 1800, *Works*, I.2.451. For the tied vote between the two republican candidates, Jefferson and Burr, in the electoral college, which was not resolved until 17 February 1801, cf. Chinard, *Jefferson*, 368–75; Peterson, *Thomas Jefferson and the New Nation*, 643–51; Elkins and McKitrick, 741–50.

on the election of Mr. Jefferson for our next President. It has been a hard struggle, but the majority for him was considerable, and much greater in the wishes of the people than in the votes of the electors. The violence of the other party, and the extremes to which they were prepared to go, are hardly credible. I myself, who had done nothing more than you will see in my Letters, was in more danger than I imagined, as I find it was under deliberation to send me out of the country under the Alien Act. It was certainly the wish of the Secretary of State, and other officers of government, but I imagined that Mr. Adams revolted at it. He desired, however, a person, who has just informed me of it, to write to me, to be upon my guard, lest I should, by Mr. —, be led to destruction.

In June he could write again to Belsham that "party-spirit is not so high as it was, owing to the moderation and prudence of Mr. Jefferson," whose letter to him, he hoped, Belsham had by now seen.[381]

"We are all republicans: we are all federalists," declared Jefferson in his Inaugural Speech of 4 March 1801—"a speech," thought Ralph Eddowes, "such as never came from the chief magistrate of any people since the world began." "If," said Jefferson,

there be any among us who wish to dissolve this union, or to change its republican form, let them stand undisturbed, as monuments of the safety with which error of opinion may be tolerated where reason is left free to combat it.

I know indeed that some honest men have feared that a republican government cannot be strong; that this government is not strong enough. But would the honest patriot, in the full tide of successful experiment abandon a government which has so far kept us free & firm on the theoretic & visionary fear that this government, the world's best hope may, by possibility, want energy to preserve itself?

I trust not. I believe this, on the contrary, the strongest government on earth.[382]

Jefferson's belief in the necessity not only of calming the party animosities which it had, erroneously, been believed were foreign to the very nature of the American Constitution, as well as encouraging the continuing presence of those whose talents and previous labours in the cause he believed to be of great practical and symbolic value for America, is clear from his correspondence. "I am in hopes you will find us returned generally to sentiments worthy of former times," he wrote to Paine on 18 March: "In these it will be your glory to have steadily laboured and with as much effect as any man living."[383] And to Priestley on 21 March Jefferson wrote similarly of his sense of Priestley's vital significance, and his embodiment of the essential philosophy of America. It was, he wrote,

---

[381] Priestley to Belsham, 2 March 1801, *Works*, I.2.454–5. The reference is surely to Thomas Cooper. Cf., however, above, n. 325; and also Priestley to Belsham, 25 June 1801, *Works*, I.2.465–6.

[382] Ford, ed., *Works of Jefferson*, IX.195–6; R. Eddowes to W. Roscoe, 4 December 1801, Roscoe Papers, L. P. L.; and see also Rush to Jefferson, 12 March 1801, *Letters of Rush*, II.831.

[383] Jefferson to Thomas Paine, 18 March 1801, Ford, ed., *Works of Jefferson*, IX.213.

with heartfelt satisfaction that, in the first moments of my public action, I can hail you with welcome to our land, tender to you the homage of it's respect & esteem, cover you under the protection of those laws which were made for the wise and good like you, and disclaim the legitimacy of that libel on legislation, which, under the form of a law, was for some time placed among them.

As the storm is now subsiding, and the horizon becoming serene, it is pleasant to consider the phenomenon with attention. We can no longer say there is nothing new under the sun. For this whole chapter in the history of man is new. The great extent of our Republic is new. Its sparse habitation is new. The mighty wave of public opinion which has rolled over it is new. But the most pleasant novelty is, it's so quickly subsiding over such an extent of surface to it's true level again. The order & good sense displayed in this recovery from delusion, and in the momentous crisis which lately arose,

Jefferson continued—referring to the constitutional crisis which had arisen in the electoral college—

really bespeak a strength of character in our nation which augurs well for the duration of our Republic; & I am much better satisfied now of it's stability than I was before it was tried. I have been, above all things, solaced by the prospect which opened on us, in the event of a non-election of a President; in which case, the federal government would have been in the situation of a clock or watch run down. There was no idea of force, nor of any occasion for it. A convention, invited by the Republican members of Congress, with the virtual President & Vice President, would have been on the ground in 8. weeks, would have repaired the Constitution where it was defective, & wound it up again. This peaceable & legitimate resource,

Jefferson concluded this "long disquisition on politics" (which was considerably more optimistic in its tone than had been some of his letters during the crisis)

to which we are in the habit of implicit obedience, superseding all appeal to force, and being always within our reach, shows a precious principle of self-preservation in our composition, till a change of circumstances shall take place, which is not within prospect at any definite period.[384]

To this disquisition on the constitutional principles of America, written as to one of her honorary citizens, Priestley very quickly replied,

---

[384] Jefferson to Priestley, 21 March 1801, Lib. Cong. Mss., Jefferson Papers, and Ford, ed., *Works of Jefferson*, IX.216–19, where "disclaim" is transcribed "disdain." For Jefferson's alarm at the potential dangers—"of anarchy" and "letting the Legislature take the nomination of the Executive entirely from the people," inherent in the deadlock, cf. his letter to Tenche Coxe, 31 December 1800, cit. Chinard, *Jefferson*, 371. Cf. also Peterson, 643–51. For Priestley's distribution of Jefferson's letter for the private perusal of his friends, "which would give you a clearer idea of him than any of his public papers," cf. Priestley to Lindsey, 8 April 1801, to Wilkinson, 30 April 1801, to Belsham, 25 June 1801, and to Hurford Stone, 19 February and n.d., 1802, *Works*, I.2.455–7, 459, 465–6, 476, and W. P. L. In 1812 Belsham, in his *Life of Lindsey*, saw fit to publish this letter of Jefferson's. Writing to Adams, who expressed some disquiet at the opinions expressed, Jefferson wrote that it had been intended as "a confidential communication of reflections . . . from one friend to another," on "the gloomy transactions of the times, the doctrines they witnessed, and the sensibilities they excited." (Jefferson to Adams, 15 June 1813, *Adams-Jefferson Letters*, II.331, and Peterson, *Jefferson*, 955–6.)

expressing his sense also of the danger in which not only he, but the whole republican experiment, had recently stood: "Had it not been for the extreme absurdity and violence of the late administration, I do not know how far these measures might not have been carried." And he wrote of his evident satisfaction in the election of a President whose political philosophy was so closely in accord with his own:

I rejoice more than I can express in the glorious reverse that has taken place, and which has secured your election. This I flatter myself will be the permanent establishment of truly republican principles in this country, and also contribute to the same desirable event in more distant ones.[385]

---

[385] Priestley to Jefferson, 10 April 1801, Lib. Cong. Mss., Jefferson Papers, and Appendix.

PRIESTLEY'S FINAL YEARS IN AMERICA
UNDER JEFFERSON, 1801–1804

In sharp contrast to the preceding years, Priestley's sphere of operations after the inauguration of Jefferson as President in 1801 was to be exclusively centered in America. For the last time, from November 1800 to the summer of 1801, he seems to have given serious thought to moving to France. In November 1800 he wrote to Russell that "as soon as there is a free and safe communication with France, I really intend to make the voyage."[386] But after his fever in Philadelphia in the spring, it was with resignation that he saw the friend whose "affection and confidence" he had once described as being his "greatest pride,"[387] depart, with all his family, for France.[388] In April Priestley could still write to the bankers in Paris who managed the funds invested for him by Wilkinson, of his intention of residing for at least part of the time in France, "if the property in France will afford me a decent subsistence."[389] But by the summer he was writing to Lindsey: "I thank you for your advice about going to France. I shall be governed by it. But really I have now very little expectation of ever seeing any part of Europe."[390]

Increasingly, the "favourable . . . turn things have taken with respect to myself" in America, as he described it to Wilkinson,[391] as much as the frailty of old age, were to render Priestley content with his lot in America. In the summer of 1800 his son Joseph had returned from England, bringing, wrote Priestley, "a very affecting account of the state of the country, such as I should think cannot continue long."[392] The climate, and even

---

[386] Priestley to Russell, 13 November 1800, *Works*, I.2.446; and cf. Priestley to Russell, 25 September 1800, *Works*, I.2.442: "To reside in France in your company, and be usefully employed there, you aiding me, as you did at Birmingham, would be the height of all my wishes. But there is a time for all things. . . ." And cf. Russell to B. Vaughan, 1 December 1800 on his intention of returning to Europe "as soon as I can embark without any danger of being taken prisoner again & it is probable Dr. Priestley may accompany me . . ." (A. P. S., Vaughan Papers, B. V. 46 p.) Cf. also Priestley to J. Vaughan, 10 November 1801 (Dickinson Colln.), for Russell's correspondence after his arrival in France: "I have received a letter from Mr. Russell at Paris, and he refers me to another that I have not yet received. I would write to him, but I wish to be informed whether there be now a free communication with France." Cf. also Priestley to Russell, 4 April 1801; to Hurford Stone, 19 February 1802, *Works*, I.2.458, 475.
[387] Priestley to Russell, 5 August 1791, 17 June 1793, *Works*, I.2.136; 202.
[388] Priestley to Russell, 4, 12 April 1801, *Works*, I.2.458-9; 459-60. Cf. S. H. Jeyes, *The Russells of Birmingham*, 277ff., for the departure of the Russells from America, and their arrival in France early in July 1801.
[389] Priestley to Citizen Perigaux, 12 April 1801, *Scientific Correspondence*, 157-8.
[390] Priestley to Lindsey, 22 July 1801, *Works*, I.2.467.
[391] Priestley to Wilkinson, 30 April 1801, W. P. L.
[392] Priestley to Lindsey, 13 August 1800, *Works*, I.2.441; Priestley to Wilkinson, 17 July, 15 December 1800, W. P. L.; and cf. Priestley to Thatcher, 10 May 1800: "I cannot help

more the government of America, he wrote to Samuel Mitchill, Professor of Chemistry at Columbia, and now a member of the House of Representatives, were both "greatly preferable. . . . Here we have *peace* and *plenty*, and in England they have neither, nor do I see that a revolution can be warded off much longer. Peace, in my opinion, will only be the beginning of internal troubles."[393] "The longer I live in this country the more I like it," he wrote to Lindsey; and he rejected, not without many expressions of gratitude, an invitation from Belsham to live with him in Hackney.[394] He continued to be much interested in the affairs of England: "The account of the debates in parliament interests me much," he wrote to Lindsey of the "Cambridge paper" which the latter still regularly sent to him, "and we have seldom anything of this in the American papers." He continued also to concern himself with the fate of those who had, with him, suffered for their exertions in the cause of liberty.[395] But he was much interested also in the details of American politics, in the state of the development of the capital city, and "what matters of importance" the members of Congress had under discussion.[396] In a long letter to Price's nephew, William Morgan, in the autumn of 1802, he described his sense of the soundness of the government and the great potential of America:

How different is our situation from yours! Our debt is trifling, and will to appearance soon be discharged, though almost all our taxes are done away. . . . There being no church establishment we have no tithe, or any expences beside voluntary ones, on account of religion, and yet there is full as much attention given to it as with you. I do not think that any country in the world was ever in a state of greater improvement, in all respects, or had fairer prospects, than this has at present.[397]

From the outset of the Administration which inspired such confidence in one who had suffered much in America, Jefferson had determined to implement the principles of sound government on which the Republicans had based their opposition to the Federalists. It was not, however, to be an Administration which was without compromise. "Some things

---

being very apprehensive for the fate of England. By accounts from my son, the scarcity there approaches to a famine. The deaths in London are more than ever was known since the great plague. All provisions are almost three times the usual prices, and yet they will not hear of *peace*. It is like that *infatuation* which, as Hartley observes, generally precedes destruction." *P. M. H. S.*, Series 2, Vol. 3 (June 1886): 37–8.

[393] Priestley to Mitchill, 16 July 1801, Schofield, *Scientific Autobiography*, 309; and cf. ibid., 365–6 for Priestley's friendship with Mitchill.

[394] Priestley to Lindsey, 13 August 1800; Priestley to Belsham, 5 June 1800; Priestley to Russell, 5 June 1800, *Works*, I.2.467; 437–8.

[395] Priestley to Lindsey, 2 October 1801, *Works*, I.2.470; and Schofield, 312. And for Priestley's interest in Richard Phillips, the radical bookseller, who had been imprisoned for his activities, cf. Priestley to Mathew Carey, 2 July 1802 (copy made in 1842): "Mr. Phillips is one of the sufferers in the case (sic: cause) of liberty in England, and one that I am on that account and various others very desirous to serve." Penn. Hist. Soc., Lea and Febiger Colln.; and cf. Priestley to Phillips, 18 June 1801, *Works*, I.2.464, expressing his gratitude for "the valuable books you have from time to time sent me."

[396] Priestley to Mitchill, 5 January 1802, Schofield, *Scientific Autobiography*, 315.

[397] Priestley to Morgan, 23 October 1802, *Works*, I.2.495–7; Schofield, 317.

may perhaps be left undone from motives of compromise for a time," Jefferson admitted in a private and very frank letter to Du Pont de Nemours early in 1802, "and not to alarm by too sudden a reformation, but with a view to be resumed at another time. I am perfectly satisfied the effect of the proceedings of this session of congress will be to consolidate the great body of well meaning citizens together, whether federal or republican heretofore called." But, as he added,

It was my destiny to come to the government when it had for several years been committed to a particular political sect, to the absolute and entire exclusion of those who were in sentiment with the body of the nation. I found the country entirely in the enemies hands. . . . When this government was first established, it was possible to have kept it going on true principles, but the contracted, English, half-lettered ideas of Hamilton, destroyed that hope in the bud. We can pay off his debt in 15 years: but we can never get rid of his financial system. It mortifies me to be strengthening principles which I deem radically vicious, but this vice is entailed on us by the first error.[398]

Jefferson's inability to rid America of the financial institutions of Hamilton, and his now declared faith in the necessity of commerce, "as the only means of disposing" of the products of agriculture, was to be a cause of much criticism. But the measures of government which in December 1801 he announced in his first Annual Message to Congress— the abolition of all internal taxation (including Hamilton's hated excise), the elimination of the national debt in fifteen years, and the reform of the self-interested Judiciary Act of the Federalists—constituted, as Duane in the *Aurora* declared, "an epitome of republican principles applied to practical purposes."[399] "War, indeed, and untoward events, may change this prospect of things," declared Jefferson, "and call for expenses which the imposts could not meet; but sound principles will not justify our taxing the industry of our fellow citizens to accumulate treasure for wars to happen we know not when, and which might not perhaps happen but from the temptations offered by that treasure." And he expressed his hope of establishing "principles and practices of administration favorable to the security of liberty and prosperity, and to reduce expenses to what is necessary for the useful purposes of government."[400]

"You have obliged me exceedingly by sending me the *President's Message*, with which I think it hardly possible for the most determined Federalist to find fault," wrote Priestley shortly afterwards to George Logan:

What a contrast does this country under the administration of Mr. Jefferson make with England under George III. It must mortify the English ministry, and I should not wonder if it be the means of bringing great numbers from that country to this. Thousands, I am confident, would come if they were able. To me the administration of Mr. Jefferson is the cause of peculiar satisfaction, as I now, for

---

[398] Jefferson to Du Pont de Nemours, 18 January 1802, Ford, ed., *Works*, IX.342-4, n.; and Chinard, *Jefferson and Du Pont de Nemours*, 35-8.
[399] *Aurora*, 18 December 1801, cit. Banning, *Jeffersonian Persuasion*, 276.
[400] Ford, ed., *Works*, IX.321-42; Peterson, *Jefferson*, 684-9.

the first time in my life (and I shall soon enter my 70th year) find myself in any degree of favour with the governor of the country in which I have lived, and I hope I shall die in the same pleasing situation.[401]

It was a measure of the mutual admiration subsisting between Priestley and the supporters of Jefferson that early in 1802 he was consulted on the measures which Jefferson had proposed in his Message to Congress. "You pay me too great a compliment by asking my opinion on the subjects you mention," Priestley in one of his more disingenuous letters declared to Logan: "I am even unacquainted with the state of the *facts*, and if I were, I am incapable of judging concerning them. This is no affectation in me. I have never given any attention to more than the great out- lines of Politicks. Further than this, my various pursuits will not admit of. All that I have heard," he nevertheless added, "concerning the new *judiciary system* left me impressed with the idea that it was not at all wanted, and that in reality nothing was meant by it but to make a per- manent provision for the friends of Mr. Adams. As to the *taxes*," he further, and rather perversely, proceeded, "I rather wish they, or the greater part of them, could have been continued, if it had been in the power of Congress to apply the produce to the farther improvement of the country." But, he added, "if no great inconvenience be foreseen to arise from it, I could wish the duty on *books and philosophical* instruments might be taken off."[402]

Priestley's admiration for Jefferson was in 1802 at its height. "He is every thing that the friends of liberty can wish," he wrote to Hurford Stone.[403] And in the summer of that year he wrote to Jefferson himself, expressing his wish to dedicate to him the second volume of his *Church History*:

It is the boast of this country to have a constitution the most favourable to politi- cal liberty, and private happiness of any in the world, and all say that it was yourself, more than any other individual, that planned and established it; and to this your conduct in various public offices, and now in the highest, gives the clearest attestation.[404]

Many have appeared the friends of the rights of man while they were subject to the power of others, and especially when they were sufferers by it; but I do not

---

[401] Priestley to Logan, 26 December 1801, Penn. Hist. Soc., Logan Papers, V.43; and cf. Priestley to Lindsey, 19 December 1801, A. P. S., Priestley Papers, B. P. 931 and Appendix.
[402] Priestley to Logan, 18 January 1802, Penn. Hist. Soc., Logan Papers, V.36.
[403] Priestley to Hurford Stone, 19 February 1802, *Works*, I.2.476.
[404] Priestley to Jefferson, 12 June 1802, Jefferson Papers. For Jefferson's correction of the implications of this passage, which, as he pointed out in his reply (19 June 1802, Ford, ed., *Works*, IX.380-2; Priestley, *Works*, I.2.483-6) was inaccurate as it stood, cf. Priestley's Dedica- tion as it eventually appeared in the second part of his *Church History*: *Works*, IX.3-6: "and all say that, besides your great merit with respect to several articles of the first importance to public liberty in the instrument itself, you have ever been one of the steadiest friends to the genuine principles and spirit of it." Cf. also Priestley to Jefferson, 29 October 1802, Appendix; and Priestley to Belsham, 30 October 1802, *Works*, I.2.498: "I fear the dedication will not suit England . . ."; and to Lindsey, 3 July 1802, *Works*, I.2.489: ". . . but I have done with that country, and am indifferent to what the friends of its government may think of me."

recollect one besides yourself who retained the same principles, and acted upon them, in a station of real power.

Jefferson's example would, he wrote, in a classic articulation of his own political philosophy,

demonstrate the practicability of truly republican principles, by the actual exis- tence of a form of government calculated to answer all the useful purposes of gov- ernment (giving equal protection to all, and leaving every man in the possession of every power that he can exercise to his own advantage, without infringing on the equal liberty of others),

and in so doing, help to render them universal. He praised Jefferson's continuing devotion to the cause of religious toleration "so that the pro- fession and practice of religion is here as free as that of philosophy, or medicine. And now the experience of more than twenty years leaves little room to doubt but that it is a state of things the most favourable to mutual candour, which is of great importance to domestic peace and good neigh- bourhood, and to the cause of all truth, religious truth least of all excepted." And he wrote again of his own consciousness of safety: "It is now only that I can say I see nothing to fear from the hand of power, the government under which I live being for the first time truly favourable to me."[405]

Priestley's now clear commitment to the government of America, the importance which his approbation represented for the leaders of her affairs, and surely, too, the influence which his political philosophy can be said to have had upon Jefferson himself, can be seen in Jefferson's reply of June 1802, expressing his pleasure that his "sincere desire to do what is right and just is viewed with candor . . . It is impossible not to be sensible that we are acting for all mankind; that circumstances denied to others, but indulged to us, have imposed on us the duty of proving what is the degree of freedom and self-government in which a society may venture to leave it's individual members." It appears too, in a letter which Jefferson wrote to Priestley in November 1802, shortly before his second Annual Message to Congress. "The quiet tract into which we are endeavoring to get," wrote Jefferson, "neither meddling with the affairs of other nations, nor with those of our fellow citizens, but letting them go on in their own way, will show itself in the statement of our affairs to Congress. We have almost nothing to propose to them but 'to let things alone.'"[406] In his Message he committed himself, perhaps to a greater degree than before, to the support of both "commerce and navigation in all their lawful enterprises, to foster our fisheries and nurseries of navi- gation and for the nurture of man, and protect the manufactures adapted to our circumstances." He pledged himself also

---

[405] Priestley to Jefferson, 12 June 1802, Jefferson Papers, and Appendix.

[406] Jefferson to Priestley, 19 June, 29 November 1802, Ford, ed., *Works of Jefferson*, IX.380-2, 404-6; and cf. also Jefferson to T. Cooper, 29 November 1802, ibid., 402-4: "A noiseless course, not meddling with the affairs of others, unattractive of notice, is a mark that society is going on in happiness. If we can prevent the government from wasting the labors of the people, under the pretence of taking care of them, they must become happy."

to preserve the faith of the nation by an exact discharge of its debts and contracts, expend the public money with the same care and economy we would practice with our own, and impose on our citizens no unnecessary burden; to keep in all things within the pale of our constitutional powers, and cherish the federal union as the only rock of safety—these, fellow-citizens are the landmarks by which we are to guide ourselves in all our proceedings.[407]

In this first year of Jefferson's Presidency, in which their mutual inspiration was made increasingly manifest, Priestley was in further correspondence with him, as a result of his own continuing communication with John Hurford Stone. In 1801 Stone had written to Priestley from France, asking for an account of the "internal administration of the United States," for the Emperor Alexander of Russia, for whose reforming ideas Stone expressed the greatest admiration, and of which he was able to give Priestley a well-informed account. The "principles . . . Sentiments and Conduct" of the reforming Emperor Stone compared to those of Jefferson. "We have now two men in the world to whom we look with mingled respect and anxiety," he wrote. His lengthy account of the current state of the affairs of France consisted however of a damning indictment of Buonaparte—"the Hero has totally disappeared beneath the Prince, and his vanity has got the better of his pride," wrote Stone. He himself, however, in spite of his contempt for the present government of France, believed that "the great principles of the Revolution" were "gaining ground every day . . . tho Buonaparte affects to do all *par* le peuple et *pour* le peuple, the people are by no means the dupe."[408]

A large part of this communication, which Priestley forwarded to Jefferson—"as the information it contains would be useful and interesting to him"—was transcribed by Thomas Cooper. It was apparently he, rather than Priestley, who was still prepared to think with Stone that "all is not lost in France." Jefferson's own hopes for the liberties of that country had for some time been at a low ebb: "The press, the only tocsin of a nation, is compleatly silenced there, and all means of a general effort taken away." "Some preparation seems necessary to qualify the body of a nation for self-government," he wrote to Priestley: "Who could have thought the French nation incapable of it?"[409] And in this it would seem that Priestley concurred. "As to Buonaparte," he wrote to Belsham in

[407] Ford, ed., *Works*, IX.406–15; and cf. also Chinard, *Jefferson*, 329–374; Beard, *Economic Origins of Jeffersonian Democracy*, 435ff.

[408] J. Hurford Stone to Priestley (n.d., 1801), copy in Jefferson Papers; and Priestley to Hurford Stone, 19 February 1802, *Works*, I.2.474: "At length I have had the satisfaction I had almost despaired of, to receive a letter from you, and one that interests me exceedingly; especially with respect to the ecclesiastical state of France and the character of the Emperor of Russia; from whom I now expect great things." Cf. also Priestley to Hurford Stone, n.d. (1802), *Works*, I.2.476.

[409] Priestley to Jefferson, 29 October 1802, Jefferson Papers, and Appendix; Priestley to Hurford Stone, n.d. (1802); Jefferson to Priestley, 29 November 1802; Jefferson to Thomas Cooper, 29 November 1802, Ford, ed., *Works*, X.402–6; Peterson, *Jefferson*, 628–9. For Jefferson's response, cf. his letter to Priestley, 29 November 1802: "Mr. Cooper's Propositions respecting the foundation of civil government; your own piece on the First principles of government . . . and the Federalist would furnish the principles of our constitution . . ."

FIGURE 19.   Joseph Priestley (1803) by Artaud (after Gilbert Stuart). Courtesy of Yale University Art Gallery. Gift of Mrs. John Fulton to the School of Medicine.

JOSEPH PRIESTLEY, LL.D. F.R.S.

AC. IMP. PETROP. R. PARIS. HOLM. TAURIN.
ITAL. HARLEM. AUREL. MED. PARIS. CANTAB.
AMERIC. ET. PHILAD. SOCIUS.

Born Mar. 13. 1733. Died Feb. 6. 1804.

FIGURE 20. Engraving of the portrait of Priestley, prefixed to his *Notes on Scripture*, published in Northumberland in 1804. By permission of the President and Council of the Royal Society.

December, "I imagine my opinion does not now differ from yours."[410] "All the friends of liberty must have been disappointed with respect to France," he wrote in one of his few extant comments on his disillusion at last with that country. "But, perhaps, a state of less political liberty may suit that nation. If they do not complain, why should we? though," he added, more honestly, "it is unpleasant to see public liberty make a retrograde motion in any part of the world."[411]

The esteem in which Priestley held the Administration of Jefferson, the proof which it happily provided of the potential for republicanism being realised within the framework of the American Constitution, and the final demise of his hopes of liberty being made secure in France, were to be reflected in the chapter (which he must at this time have been writing) on the Constitution of America, for incorporation into the revised edition of his *Lectures on History and General Policy*, published in Philadelphia in 1803. The Revolution in France, he now went so far as to admit, had demonstrated how difficult it was to predict the outcome of changes in government: "The system established" there "at present is the very reverse of every thing that was intended at the commencement of the revolution."[412] His critique of the American Constitution was far removed from the acerbity of his *Letters* of 1799. And there is to be found in it, too, a renewed stress upon the value of individual responsibility, and an awareness of the danger of the interference of government, the specific emphasis of which can almost certainly be ascribed to his close propinquity to Jefferson, and his growing confidence in his Administration.

The essential equality of all men, the realisation of their natural rights within the framework of the interests of the whole, and the need to render government simple, efficient, and inexpensive were, as they had always been, the cornerstones of Priestley's thought. They had also been central to the philosophy of the *Rights of Man*. But whereas Paine, in *Agrarian Justice*, had in 1797 called for a recognition of the need for government to redress the inevitable inequalities of society, for Priestley in 1803, as for all good Jeffersonians, the best interests of the individual could only be served by a minimum of such interference. "The great excellence of this constitution," he wrote,

consists in the simplicity of its object, which is the security of each individual in the enjoyment of his natural rights, without aiming at much positive advantage; by which means every person knowing that he will be effectually protected from violence and injustice, both against the evil-minded of his fellow-citizens,

---

[410] Priestley to Belsham, 18 December 1802, *Works*, I.2.500; and cf. Priestley to Russell, 10 April 1800: "I do not know what to think of the new Constitution of France"; and Priestley to Belsham, 30 March 1800, *Works*, I.2.430, 431; Priestley to Thatcher, 10 May 1800, *P. M. H. S.*, Series 2, Vol. 3 (June 1886): 37–8.

[411] Priestley to William Morgan, 23 October 1802, *Works*, I.2.497; and Schofield, *Scientific Autobiography*, 317–18.

[412] Priestley, *Lectures on History and General Policy; to which is prefixed, an Essay on a Course of Liberal Education for Civil and Active Life: and an Additional Lecture on the Constitution of the United States* (Philadelphia, 1803), *Works*, XXIV.35, note.

and the enemies of his nation, will be at full liberty to employ all his faculties for his own advantage; and this he will better understand, and provide for, than the state could do for him.

The power of the whole community may be easily united in works of acknowledged public utility, as roads, bridges, and navigable canals, and also in providing the means of education, of which all the citizens may take advantage.

The history of all the European governments shews that there is no wisdom in any government aiming at more than this. . . .

If governments attempted to ordain anything more for their subjects, it would be a self-defeating endeavour: it "would be an effectual stop to all improvements. For every improvement, being suggested by individuals, would be opposed by the more ignorant and bigoted majority, educated in the old imperfect methods."[413]

For Priestley, as for so many of the philosophers of the age, the faith to which they had so long adhered, of the essential wisdom of the rule of the majority, was to be severely tempered by experience. In his discussion of the American Constitution, this change of emphasis was to be very clear. He defended a system of delegated political power:

The mode of choosing by electors leaves the choice to those who are better qualified to judge than the greater number who choose them. At the same time the electors, being few, are under a greater degree of responsibility. All history shews that the more numerous is the body that decides upon any thing, the more hasty, intemperate, and injudicious are their resolutions. In a multitude they are but few who really think and judge for themselves: consequently they are guided by a few who do think; but being under no particular responsibility, are often influenced by their private views to mislead the rest.[414]

He was now prepared to admit—in sharp contrast to his *Letters* of 1799—that the method of election, the composition, and the powers of the Senate might be an advantage: "A set of men of greater age, and experience, not chosen by the common people, and who continue a considerable time in office," could, he wrote, be "a check" upon those who, elected at shorter periods, might be led into "hasty and improper resolutions." He agreed also that the distribution of the Senators—two for every state—was an advisable arrangement. And he believed further that for the members of the House of Representatives to be elected every two years, instead of one, was a useful check against an excess of popular influence.[415] He still, however, criticised the extension by the President and the Senate of their power of making treaties: as, he wrote,

nothing is said of the limitation of that power, they have claimed, and exercised, the power of making treaties to regulate commerce, a power which is expressly confined to the whole congress; and on the same pretence they might make

---

[413] Priestley, *Lectures, Works*, XXIV.255; and cf. especially Bonwick, "Joseph Priestley, Emigrant and Jeffersonian," 17–21; Kramnick, "Priestley's Scientific Liberalism," 17ff.; "Republican Revisionism Revisited," 645–6; and "Revolutionary Philosopher, Parts I and II."
[414] Priestley, *Lectures, Works*, XXIV.255.
[415] Ibid., 255–6.

treaties offensive and defensive with foreign nations, and thus involve the country in a war.[416]

He also pointed out, although considerably less forcefully than in 1799, the opportunity for corruption in a Presidential office which could, by constant renewal, be held for life. "On the other hand," he added, with, again, a striking change of emphasis from his former exclusive dependence upon the wisdom of the popular voice, "there is a disadvantage in frequent changes of the president, on account of a possible change of general maxims and views in government."[417]

The sense of security so evident in Priestley's critique of the American Constitution in 1803, and his accord with Jeffersonian principles of government, are evident also in his concluding remarks. It was, he wrote, sound policy in America that men in public office should be so meagerly rewarded. This prevented corruption, and there was no lack of persons "of independent fortunes" capable of serving the country; nor were those not so endowed necessarily excluded from the political process:

In the present state of things, men of talents, but without fortune, may think themselves happy in a country the government of which is so excellently constituted, and so peculiarly favourable to ingenuity and industry, by means of which they may serve themselves, and the country too, in many ways, independently of having access to public offices. They are not prevented from suggesting hints to those who do act, though they cannot act themselves. To annex certain privileges to the acquisition of property,

Priestley concluded, "operates as a motive to industry, by which property may be acquired, and this ought to be encouraged by the laws of every country." And he finished his chapter by re-stating the principles which he had enunciated in *Political Arithmetic*:

what appears to me to be of particular importance as *a maxim of policy*, in the present state of the country in general . . . not to favour one class of the citizens more than another by any measure of government, especially the merchant more than the farmer. Their employments are equally useful to the country, and therefore they are equally entitled to attention and protection, but not one more than the other. If the merchant will risk his property at sea, let him calculate that risk, and abide by the consequence of it, as the husbandman must do with respect to the seed that he commits to the earth; and let not the country consider itself as under any obligation to indemnify one for his risks and losses any more than the other, especially as, in the case of the merchant, it might be the cause of a war with foreign states.[418]

"I trust that *Politics* will not make you forget what is due to *science*," Priestley had written to Jefferson in the spring of 1801.[419] But after his own involvement in the tumult surrounding the presidential campaign,

---

[416] Ibid., 256–7.
[417] Ibid., 257.
[418] Ibid., 258–9.
[419] Priestley to Jefferson, 10 April 1801, Jefferson Papers and Appendix.

his interest in the development of the country with which he now clearly identified as most perfectly embodying his republican ideal of individual and collective liberty, was undoubtedly very great. In this he was joined by the fiery extremist who was, in his last years, increasingly in his company, and whose developing political career in America Priestley watched with some admiration. "I *am satisfied* that of republicanism you know little or nothing in Europe," wrote Thomas Cooper after Jefferson's election, to young Watt, in a letter which graphically described his own trials in the cause:

& that if the experiment is to be made at all, it must be made here, & if here, I know no administration but that which is likely to take place under Jefferson which will give it fair play. But I see the manifold advantage of the System & altho' I am not like that literary egotist Godwin a Champion for the perfectibility of the human race, yet I am perfectly satisfied that the utmost amelioration of the condition of mankind which even sanguine philosophy can expect, must arise from the principles of republican Government.[420]

The satisfaction which both Priestley and Cooper took in the success of the American experiment in government, can be seen in one of Priestley's last letters to Belsham, written in the summer of 1803, as Europe to his distress became embroiled once more in the misery of war. "We are happily at *peace* here," he wrote, "and without the most distant prospect of war." He described to his friend the acquisition of Louisiana — the affair on which "every eye in the United States" had been fixed:

The opposition was clamorous for taking possession of New Orleans by force; but now that, and all Louisiana, is to be gained without it, and in a manner much more likely to be permanent. Had it remained in the possession of France, it

---

[420] T. Cooper to J. Watt, Jr., 1 February 1801, B. R. L.: "Having staid in the country during six years without engaging in politics, accident made me become the editor of a Newspaper for a short period to accommodate Dr. Priestley, & since that time I have been plunged up to the ears in my old pursuit. The obloquy and the popularity I have experienced here during this short period has far exceeded anything of the kind I was exposed to in England. I have been imprisoned 6 Months for a Libel on the President, I have been fined about 700£ sterling, which the public took good care I should not pay, & the party whom very conscientiously I espoused & eminently served, is now coming into power, and I have every reasonable expectation that my Situation will be respectably mended." Cf. *An Account of the Trial of Thomas Cooper* . . . (Philadelphia, 1800); Malone, *Cooper*, 134ff.; Smith, *Freedom's Fetters*, 317–33, for Cooper's trial, attended by "most of the high-ranking Federalists." Cf. also *Cyclopaedia of American Literature*, II.333, for Cooper's later reminiscences of the good company which he enjoyed, "every day and night," while serving his six-month prison sentence in the latter half of 1800: "At night I had the best company in Philadelphia. They all called on me." Cf. also Priestley to Lindsey, 29 May 1800, *Works*, I.2.436: "Mr. Cooper has been convicted of a libel, on the Sedition Act, and is now in prison; but he has gained great credit by it, and he will, I doubt not, be a rising man in the country. The trial is published, and I shall send you a copy of it." And Priestley to Hurford Stone, n.d., 1802, *Works*, I.2.476: "He makes himself conspicuous, and will be a rising man in this country." Cf. also Priestley to Barton, 27 November 1800, Schofield, *Scientific Autobiography*, 307–8, for his concern when Cooper was rejected by the American Philosophical Society: "Can you explain it to me? I fear that *Politics* had something to do in it. . . ." For Cooper's election to the Philosophical Society in January 1802, cf. Malone, 170–1.

would no doubt have been taken from them by the English, and they would have completely inclosed all the United States to the west. But the increasing population of this country would in time have burst through that feeble barrier.[421]

"We rejoice in the success of your mission by the peaceable acquisition of Louisiana," he wrote to Livingston, the principal negotiator of the purchase: "How much more satisfactory, as well as more secure, is an acquisition made in this manner, than by war." And the last letter which he wrote to Jefferson was to congratulate him on this signal triumph for American diplomacy: "When you wrote to me at the commencement of your administration you said 'the only dark speck in our horizon is Louisiana.' By your excellent conduct it is now the brightest we have to look to."[422] The danger had indeed been very great, Jefferson replied. But, he added, "the *dénouement* has been happy; and I confess I look to this duplication of area for the extending a government so free and economical as ours, as a great achievement to the mass of happiness which is to ensue." And he asked Priestley whether he had read the "new work of Malthus on population," of which he himself had only been able to see a borrowed copy. "It is one of the ablest I have ever seen . . . probably our friends in England will think of you, & give you an opportunity of reading it."[423]

To an extent which had never been properly recognised in England, Priestley's achievements, his acuteness, grasp and prophetic vision as a political philosopher, and "the vast exertions of his genius," as Jefferson was many years later to describe it[424] were, if belatedly, given their due in America. "The notice of me which you are so good as to prefix to your book," Jefferson had written to Priestley in 1802, on the dedication to him of the second volume of the *Church History*, "cannot but be consolatory, inasmuch as it testifies what one great and good man thinks of me."[425] Early in March 1803, on what was his last visit to Philadelphia, the American Philosophical Society gave a dinner as "a testimony of respect, to their venerable associate Dr. Joseph Priestley."[426]

By the summer of 1803, Priestley was writing of his growing sense of decline. He had, he wrote to Benjamin Smith Barton, "nothing to communicate" to the *Transactions* of the Philosophical Society. "Indeed," he wrote, in a phrase very strikingly reverting to memories of his early

---

[421] Priestley to Belsham, 6 August 1803; Priestley to Lindsey, 11 July 1803, *Works*, I.2.513–16; Peterson, *Jefferson*, 753.

[422] Priestley to Livingston, 12 December 1803, Charles Roberts Autograph Colln., Haverford College. Priestley to Jefferson, 12 December 1803, Jefferson Papers. (It was in fact in Jefferson's letter of 29 November 1802 [above, n. 406], that he had mentioned Louisiana.)

[423] Jefferson to Priestley, 29 January 1804, Ford, ed., *Works*, X.70–2; Rutt, ed., *Works*, I.2.524–5.

[424] Jefferson to Adams, 22 September 1813, L. J. Cappon, ed., *The Adams-Jefferson Letters*, II.378.

[425] Jefferson to Priestley, 29 November 1802.

[426] Rush to Jefferson, 12 March 1803, *Letters*, II.858. Schofield, *Scientific Autobiography*, 318–19. And cf. Priestley, *Memoirs* I.170 for the offer by the University of Pennsylvania to Priestley of the office of Principal—an offer which he declined.

forays in politics in England, "I am, as Mr. Wilks used to say, *an exhausted volcano.*" And in another of his last surviving letters, written to Logan at the end of January 1804, he forecast accurately, and with his usual dispassionate composure, his own imminent demise. "Tell Mr. Jefferson," he wrote, "that I think myself happy to have lived so long under his excellent administration; and that I have a prospect of dying in it. It is, I am confident, the best on the face of the earth, and yet I hope to rise to some thing more excellent still."[427] On 6 February 1804 Priestley, attended by his son Joseph, his daughter-in-law Elizabeth Ryland, and Thomas Cooper, died in his home in Northumberland. He was working upon his annotations to the Old and New Testaments to the last; and "he dwelt," as his son later wrote, "upon the peculiarly happy situation in which it had pleased the Divine Being to place him in life; and the great advantage he had enjoyed in the acquaintance and friendship of some of the best and wisest men in the age in which he lived, and the satisfaction he derived from having led an useful as well as a happy life."[428] To his scientific friends in Philadelphia, Benjamin Rush, Benjamin Smith Barton, and James Woodhouse, Thomas Cooper on the day of Priestley's death wrote to inform them of the news.[429] It was almost certainly Thomas Cooper, also, who on the same day composed a tribute to Priestley for publication in the *Aurora*, in whose columns it appeared shortly afterwards. It dwelt upon Priestley's achievements as a proponent of rational Christianity; on his high standing as a metaphysician; and on his contribution towards the studies of belles lettres and chemistry. And "as a politician," this notice of Priestley's life recorded,

he has assiduously and successfully laboured, not merely to prepare the minds of his former countrymen of Great Britain, to adopt those gradual and salutary reforms in their own system of government, which the democratic part of it so obviously requires, but to extend and illustrate those general principles of civil liberty which are happily the foundation of the constitution of his adopted country.[430]

In another notice of respect in Philadelphia, extracted from a Northumberland newspaper, came a similar tribute to Priestley's great achievements in science, religion, and philosophy. Nor, it added, "ought it to be forgotten that he spent much of his valuable time, and to a purpose no less valuable

---

427 Priestley to B. S. Barton, 12 July 1803, Schofield, 319–20; Priestley to Logan, 25 January 1804, Penn. Hist. Soc., Barton Papers, p. 65.

428 Priestley, *Memoirs*, I.216–17.

429 Cooper to Benjamin Rush, 6 February 1804, *Scientific Correspondence*, 162; Cooper to Benjamin Smith Barton, 6 February 1804, Schofield, *Scientific Autobiography*, 321; and cf. ibid., 322, for the notice of the Memorial Service held by the American Philosophical Society on 29 December 1804, at which the Eulogy for Priestley was delivered by Benjamin Smith Barton. Cooper's letter to James Woodhouse, the professor of Chemistry at the University of Pennsylvania, was published in the *Aurora* in part on 15 February 1804: cf. below, Appendix.

430 *Aurora*, 15 February 1804; and cf. ibid., 18 February 1804, for a detailed account of the circumstances of Priestley's death.

in asserting and establishing those principles of civil liberty which America hath so wisely adopted and so happily put into practice."[431]

From Maine Benjamin Vaughan, in a letter to his brother John, wrote that "with sincere affliction" he had learned of "the death of Dr. Priestley. I looked upon him," wrote Vaughan—all differences between himself and his former mentor now happily forgotten—"as our relation; a close acquaintance of near 40 years, with mutual confidence, having intitled me to do this. But I shall not," he added,

choose to write any for the New England papers. It ought rather to appear to the Southward, and travel hither; when it will be fathered upon no particular individual. I am however doubtful, whether it will not be better to leave the matter to others; & reserve myself for other modes & times.

Priestley's "*life*," he prophetically remarked, "perhaps can be written by no one, with ease, now he is himself gone."[432] In 1806 Priestley's *Memoirs* were published, continued by his son Joseph, and with contributions from Thomas Cooper and William Christie. With the exception of the remarks in Thomas Cooper's Essay, however, they revealed no more than Priestley himself in his printed record of his life had been prepared to admit, of his interest in politics. And it was almost certainly while preparing them that Joseph Priestley the younger destroyed much evidence which might have suggested the contrary—all the correspondence received by his father while in America.[433]

In America Priestley's memory was to be well served by Thomas Cooper, whose own intimacy with Jefferson was to exceed that of the older man.[434] It was to Cooper that Jefferson wrote in 1807, on hearing of the publication of Priestley's *Memoirs*, that, had he known he would "certainly have procured it; for no man living had a more affectionate respect for him. In religion, in politics, in physics," wrote Jefferson, "no man has rendered more service."[435] It is in the correspondence between Jefferson and Adams, however, two of his oldest surviving political and philosophical acquaintance, that Priestley's name most frequently recurs. "I never recollect Dr. Priestley, but with tenderness of Sentiment. Certainly one of the greatest Men in the World, and certainly one of the weakest," wrote Adams to John Vaughan in 1813.[436] To one English

---

[431] *Poulson's American Daily Advertizer*, 16 February 1804.

[432] B. Vaughan to J. Vaughan, 1 March 1804, Vaughan Papers, A. P. S., B. V. 46: "However," he added, "if you know of particulars, especially as to his death, & as to his life &c. as far as spent in Pennsylvania, it will be well to send them." Cf. also B. Vaughan to J. Vaughan, 12 November 1804, ibid., in which he appears to be sending information on the details of Priestley's life for the Memorial Service Eulogy at the Philosophical Society (above, n. 429).

[433] Schofield, *Scientific Autobiography*, ix.

[434] Malone, *Cooper*, 226; 391; and cf. ibid., 165-73; 222-8, for Cooper's close relationship with Jefferson throughout his subsequent career in America.

[435] Jefferson to T. Cooper, 9 July 1807; and cf. also his subsequent letter to Cooper, 1 September 1807: "I shall read (the *Memoirs*) with great pleasure, as I revered the character of no man living more than his." Ford, ed., *Works of Jefferson*, X.450-3.

[436] Adams to J. Vaughan, 23 November 1813, A. P. S., B. V. 462.

visitor Adams confided his belief that Priestley's intervention had in all probability contributed to his defeat in the election of 1800.[437] It was to Jefferson however that he wrote: "Oh! that Priestley could live again!" And it was to one who, with Priestley, had parted company with him in the political controversies that prevailed in America in the 1790s, that Adams confided his characteristically trenchant estimate of Priestley's leading role in the thought of the age:

The fundamental Principle of all Phylosophy and all Christianity is "REJOICE ALWAYS IN ALL THINGS. Be thankfull at all times for all good and all that We call evil." Will it not follow, that I ought to rejoice and be thankful that Priestley has lived? Aye! that Voltaire has lived?[438]

---

[437] R. B. Davis, ed., *Jeffersonian America* (Huntington Library, California, 1954), 330.
[438] Adams to Jefferson, 3, 25 December 1813, *Adams-Jefferson Letters*, II.405, 409-10.

CONCLUSION

It was as "the quintessential English philosophe," who, "more than anyone else . . . qualifies as the central intellectual figure" among the radicals of his generation, that Isaac Kramnick rightly characterised Priestley's standing in the political and intellectual life of England before his departure for America in 1794.[439] Priestley's stature and the influence which he wielded was implicit in the Address sent to him on his emigration by the United Irishmen, and in Coleridge's famous apostrophe: "Patriot, and saint, and sage."[440] "The emigration of Dr. Priestley will form a striking historical fact," wrote the United Irishmen, "by which alone future ages will learn to estimate truly the temper of the present times. . . . But be cheerful, dear Sir; you are going to a happier world. . . . Soon may you embrace your sons on the American shore, and Washington take you by the hand, and the shade of Franklin look down with calm delight on the first statesman of the age extending his protection to its first philosopher."[441] The United Irishmen could not foresee the chequered fortunes that lay in store for Priestley in America. But they rightly foretold the great impact his arrival there would have, the recognition which, to a greater degree than in England, he was eventually to be accorded, and the symbolic significance which his presence in America represented for those who saw in the election of 1800 the triumph of republican principles over the advocates of "kingly government." Only the "unyielding opposition of those firm spirits who sternly maintained their posts," wrote Jefferson many years later, had rescued American republicanism from the danger of extinction.[442] It was as to one of those "firm spirits" that he paid tribute to Priestley in March 1801, and acknowledged his great symbolic significance for America:

Yours is one of the few lives precious to mankind, & for the continuance of which every thinking man is solicitous. Bigots may be an exception. What an effort, my dear Sir, of bigotry in Politics & Religion have we gone through! . . . But it was the Lilliputians upon Gulliver. Our countrymen have recovered from the alarm into which art & industry had thrown them; science & honesty are replaced on their high ground; and you, my dear Sir, as their great apostle, are on it's pinnacle.[443]

---

[439] Kramnick, "Joseph Priestley's Scientific Liberalism," 30; "Republican Revisionism Revisited," 645.

[440] S. T. Coleridge, "Religious Musings," in E. H. Coleridge, ed., *The Complete Poetical Works of Samuel Taylor Coleridge* (Oxford, 1912, repr. 1962), I.123.

[441] Priestley, *Works*, I.2.218–22.

[442] Ford, ed., *Works of Jefferson*, I.167–83; and Banning, *The Jeffersonian Persuasion*, 13–14.

[443] Jefferson to Priestley, 21 March 1801, Ford, ed., *Works of Jefferson*, X.216–19.

From the time of Priestley's arrival in America in 1794, the developing political tension in that country had been reflected in the outpouring of enthusiasm with which he was initially greeted, and in the growing hostility which his presence aroused. As a political refugee from the turmoil of Europe, and as a committed and influential exponent of republicanism, Priestley was to become involved in the violently heated debate in America as to its viability as a practical form of government. The strongly held conviction of Hamilton and John Adams of the very possible need for the re-establishment of monarchical forms of government in America, to prevent the de-stabilisation they feared was proving inherent in democratic government, influenced much of the thinking and policies of Adams's Administration. And an awareness of the reality of this threat to the republican ideal fueled the efforts of the "anti-federalists" in their crusade against the measures of the "monocrats." Seen in this light, the victory of the republicans in 1800 was, as Jefferson in his famous Inaugural, and his supporters throughout the country, declared, the successful vindication of the entire republican experiment: "Some times it is said that Man cannot be trusted with the government of himself—Can he then be trusted with the government of others? Or have we found angels in the form of kings to govern him?—Let History answer this question." With all the natural blessings which America enjoyed, said Jefferson, what more was

necessary to make us a happy and a prosperous people? Still one thing more, fellow citizens, a wise & frugal government, which shall restrain men from injuring one another, shall leave them otherwise free to regulate their own pursuits of industry & improvement, and shall not take from the mouth of labor the bread it has earned. This,

declared Jefferson, "is the sum of good government, & this is necessary to close the circle of our felicities."[444]

Jefferson's Inaugural of 1801 was greeted with enthusiasm bordering upon veneration by more than one English radical. By none, however, was it welcomed more than by the two radical intellectuals who had supplied the republicans with valuable propaganda, and who fully endorsed Jefferson's interpretation of the true measure of his victory in 1800. In Thomas Cooper's letter to James Watt, Junior, in February 1801,[445] and in Priestley's frequent encomiums on the development of republican government under Jefferson, there is ample evidence of the sympathy with which they viewed Jefferson's policies, and of the degree to which they believed that in America alone were the experiments in government which had so preoccupied the most active minds of their gen-

---

[444] Ford, ed., *Works*, IX.193–200; and cf. Banning, *Jeffersonian Persuasion*, 270, for the editorial of 20 February 1801 in the *Aurora*: "On Tuesday last" (i.e. 18 February) . . . the question . . . on the issue of which rested the liberty, Constitution and happiness of America was terminated as every republican and honest man wished. . . . On that day the sun of aristocracy set, to rise no more." And cf. also ibid., 302.

[445] Cf. above, n. 420.

eration capable of implementation. Their views on some issues of social and economic policy had been considerably modified from those which they had held in the very different circumstances of England. Their fundamental democratic stance, however, remained unaltered, and was articulated in their commentaries on the critical issues at stake in the election campaign of 1800–1801.

It was above all for its practical implementation of the principle of the sovereignty of the people in the affairs of government that Priestley and Thomas Cooper held dear the government of America. Priestley, wrote Thomas Cooper in his account of the political works of the latter in his *Memoirs*, had ably expounded the principles of civil government in his *Essay* of 1768 – and it had "the more merit, as the experiments on government since made in America, had not then been thought of." Priestley had, however, wrote Cooper,

the satisfaction to live long enough to see a government whose theory was in his opinion near perfection, administered under the auspices of his friend Mr. Jefferson in a manner that no republican could disapprove. To the end of his days, this was a source of great satisfaction to him, especially as it became more and more evident from the disorders attendant on the French revolution, that if the republican system was required to stand the test of experiment, it was in this country alone, and under such an administration as he witnessed, that it stood any chance of success. At present,

concluded Cooper, "the trial justifies the anxious hopes of its supporters, and bids fair to establish beyond all doubt, the superiority of that form of government, on which the political reformers of modern days have rested their most reasonable expectations, and their fondest hopes."[446]

It was as one to whom "the political reformers of modern days" owed much, but who had suffered greatly for his dedication to a cause in which he profoundly believed, that Priestley in 1794 came to America. If his years in this country were to test once again his fundamental democratic vision, it was nevertheless in America that he lived to see his hopes fulfilled.

---

[446] Priestley, *Memoirs*, II.354–5; 366–7.

APPENDIX

John Vaughan to Joseph Priestley: 3 October 1791
[A. P. S., Vaughan Papers, B. V. 462.1]

Philada. 3 Oct 1791

Revd. Dr. Joseph Priestly
Birmingham
Dear Sir
     This additional mark of your confidence in our funds has given Singular pleasure to those in the public departments they have learnt with great mortification that the conduct & principles they So much honor & esteem should have subjected you to So Severe and Cruel an attack as you have experienced; if these persons whose esteem arises only from the reputation you hold are thus affected, what must be the feelings of one whose heart is animated with the warmest friendship. After your prudent & liberal address to the people of Birmingham I must check the violence of expression which Such conduct would naturally have provoked. The people in power will I hope be Sufficiently alarmed at the effects of their Ill judged encouragement of illiberal & unmanly Sentiments; if they are not & persist in a Similar line of conduct I have no doubt they will have to lament the loss & we to falicilitate (sic) ourselves upon the accession of a considerable number of the most enlightened liberal & industrious of her Citizens, could this have been obtained by us on other terms we should have been happier; if obtained upon these terms they will have a double title to a favorable reception & will meet with every encouragement in our power to bestow.
     I remain &c.
     J. V.

John Adams to Joseph Priestley: 19 February 1792
[Mass. Hist. Soc., Adams Papers, Reel 115]

Philadelphia
Feb 19th 1792

Dear Sir,
I take an opportunity by part of my family bound to London, to remind you of a person who once had an opportunity of knowing you personally, and to express my sympathy with you under your Sufferings in the cause of Liberty. Inquisitions and Despotisms are not alone in persecuting Philosophers. The people themselves we see, are capable of

persecuting a Priestly, as another people formerly persecuted a Socrates. By a compliment which I hold very precious in your familiar letters to the Inhabitants of Birmingham, I am emboldened to hope that you will not be displeased to receive an other Coppy (sic) of my Defence, especially as that which was presented you formerly has probably had the honor to share the fate of your Library. As there is not a Sett to be sold in London . . . I have desired Col. Smith to take one from New York, and present it to you with my sincere veneration.

This Country is as happy I believe as it ever was, or will be. Ambition and Avarice however, exist here, as well as in England, and produce contests and dissentions, their usual fruit. The office of President, with its twenty five thousand Dollars, will glitter in the eyes of Americans, very nearly as much as that of King in England with his Millions. I am Sir with sincere sentiments of esteem, and Respect, your most obedient

John Adams

The Revd Dr. Priestly
London

Joseph Priestley to John Adams: 20 December 1792
[Mass. Hist. Soc., Adams Papers, Reel 375]

John Adams Esq
  Vice-President
  Philadelphia
  via New York
    Dear Sir,

      I feel myself much gratified, and highly honoured, by the sympathy which you express with me on account of my sufferings in the riot at Birmingham. The same malignant spirit, fostered by our governors, is much more prevalent now than it was then, and shows itself in almost every part of the kingdom, so that I begin to fear the most serious evils from it. Nothing has yet been done towards our indemnification, tho a year and half are now almost elapsed since the event, and it is said that the officers dare not collect the little that was awarded us.

Many Dissenters wish to leave a country in which they find neither protection nor redress; but they are at a loss where to go, and how to proceed. Yesterday I received a letter from a great number of Dissenters in the neighbourhood of Manchester to Mr. Vaughan, desiring his advice in the business, and yours would be considered as a very great favour. France being in an unsettled state, I think it very probable that some of my sons will be disposed to go to America; and if so, I shall follow them in due time.

A war with France is much talked of, but I cannot think that our court, tho ever so willing, will risk such a measure. That must soon bring our affairs to a crisis.

I thank you for your very acceptable present of your three volumes, two of which were destroyed in the riot. It is a work of great value, tho I cannot say but I now think more favourably of a pure republic than I have done. A comparison between the American and French governments some years hence will enable us to form a better judgment than we can at present.

We must not expect that Ambition or Avarice will ever cease to influence mankind, but certainly there are fewer objects of those passions with you than with us, and therefore they cannot produce so much mischief. But indeed I am no politician, and I would gladly confine myself to my theological and philosophical pursuits, if I might be permitted (torn: ?so) to do.

> With the greatest gratitude, and respect, I am,
> > Dear Sir,
> > > yours sincerely

Clapton. Dec. 20 . 1792   J. Priestley
Endorsed: "Dr Priestley. Dec. 20 1792
> ansd. Feb. 27. 1793."

Joseph Barnes to Thomas Jefferson: 17 August 1793
[Lib. Cong. Mss., Jefferson Papers]

(Extract)
> > London Augt. 17th 1793

Sir

This will be presented to you by Mr. Priestley, son of the celebrated Doctor Priestley, who goes to the United States to Seek an Asylum for his father, And, who, previous to Making a purchase, Means to visit all those parts of the States which he conceives an object, in order to enable him to determine on the most eligible place to reside—

I am happy in giving him this Letter to you, not only, from a knowledge of your being the Most competent to advise, but from a full Sense of the pleasure you will receive in giving; And he in receiving your advice on the Object of his Mission.

> With the highest esteem
> I am Sir
> yours most respectfully,
> Joseph Barnes

Mr. T. Jefferson

Joseph Barnes to Thomas Jefferson: 17 August 1793
[Lib. Cong. Mss., Jefferson Papers]

(Extract)
> > London Augt. 17th 1793

Sir

This will be presented to you by Mr. Cooper, of Manchester, who is concerned in one of the principal cotton Manufacturys in that place. And, who, from his great efforts in Society, and in writing in favor of the specific rights and general Liberty of Mankind, has become so offensive to the present Spirit of the British Government, that he can no longer in Safety reside in this country; he therefore goes to Seek an Asylum in the United States.

As I esteem you, Sir, our great Patron of Republicanism, and of Virtue, 'tis with peculiar pleasure I give him this letter to you,—Your Patronage to him will follow of course, so far as you May find him worthy thereof, And I am sure he will wish it no farther.

    With grateful esteem
      I am Sir
      Yours most respectfully
      Joseph Barnes
Mr. Jefferson

P.S. The Book which Mr. Cooper has written & Published, I have been well informed, contains the essentials of Mr. Paine's Rights of man, in excellent Language & great demonstration—he is a friend of Doctor Priestley & of Mr. Walker of Manchester. J.B.

Joseph Priestley to John Vaughan: 6 February 1793
[A. P. S., Priestley Papers, B. P. 931]

Clapton Feb 6. 1793

Dear Sir

I cannot express how much I think myself obliged to you for your kind offer in your last letter, tho I do not yet know how to avail myself of it. Such is the increasing bigotry and violence of the High church party in this country, that all my sons must leave it, and settle either in France or America. As my daughter, however, must remain here, I own I should incline to France, which is so much nearer, if that country was settled. But I fear it is far from being so, and the war that has in a manner commenced between England and France will add much to the difficulty of going thither. My son Joseph, who was settled at Manchester, is under a necessity of leaving it, and as things are situated, he inclines to America, and proposes to spend a year in looking about him. Many others, at least a hundred families, will also leave that neighbourhood; but they think of going to Kentucky. As I shall, in all probability, follow my sons, I incline to the neighbourhood of *Boston*, where, I imagine, the society will suit me best, and I should like to be within a moderate distance of my sons. Their going, and settlement, must precede mine, and as I cannot bring my Apparatus with me, so as to do much after my removal, I own, I am willing to defer it as long as I can. We seem to have a very melancholy prospect before us, and perhaps it may not be easy even to get to America, as the sea will swarm with French privateers.

I thank you for the account you sent me of my interest in your funds, but you did not include the thousand £ that your brother Charles bought. As I do not write to him, I wish you would desire him to write to his brother here about it.

You are happy in being free from the alarms of *riots*, or of *war*, and also from the influence of *church power*. Here the dread of republican Principles that has seized our governors drives them to numberless acts of cruelty. Prosecutions for seditious publications, and speeches, are without number, and the late Act against *Aliens* executed with great rigour to the distress of many innocent and worthy persons. We hope, however, that this extreme violence will not last long.

> With much respect, I am,
>> Dear Sir,
>>> yours sincerely,
>>> J Priestley

P.S. I sent my son Wm a copy of your kind letter but I have not found that he has received it. All letters to or from France are opened.

Addressed: "John Vaughan Esq, Philadelphia."
Endorsed: "Joseph Priestley
    Clapton Feby 6th 1793
    Recd. April 9th 1793."

Joseph Priestley to John Vaughan: 8 June 1794
[A. P. S., Misc. Mss.]

New York June 8th 1794

Dear Sir.

I thank you for your very friendly letter, wch. I have just received; but be assured that the conduct you wish me to pursue is the very same I had prescribed to myself before I left England, & my conduct there will shew that I have been thro' life as little as can well be supposed of a political character, having only been an advocate for general liberty, & a free representation of the people as the foundation of it. As for the political parties of this country, I had not so much as heard the names of them, & I am not disposed to make much enquiry about them.

My opinion concerning clubs & political associations in general is this. If their object be to promote political discussion, they are highly useful; but if they tend to bind men to any species of political conduct, besides promoting curiosity, they are an obstruction to real freedom of thinking and acting. However, a wise man will never take much umbrage at them. If, undesigned, as I believe is generally the case, by the authors and promoters of them, they should lead to any mischief, it will soon be apparent, & by the influence of sensible & moderate men it will be prevented. And in the mean time, whatever be the design of the authors of them, they will only promote useful discussion, tho' perhaps more acrimonious than might be wished.

In a government so fundamentally good as that of this country, a government, as Dr. Price observed without Bishops (meaning such as those in England) without nobles & without a king (meaning an hereditary chief governor) whatever be imperfect, & requires amendment, will, I doubt not, in due time find it. As to myself, I have seen & felt so much of the greater abuses of government, that I shall perhaps be even too little attentive to smaller ones. For these ought to be narrowly watched, lest they should lead to greater.

Lest any person should have been led to mistake my views & principles, I have no objection to your making this letter as public as you please.

> I am, Dear Sir
>> Yours Sincerely
>> J. Priestley

P.S. I have seen reason to think that the publication of this letter to you (in which you may mention your own name, or not, as you think proper) in the Pensilvania News Paper, and then in that of New York, will be of material Service to me; much pains having been taken, I am told, by emissaries from England, to represent me as a (?) and dangerous person.

I hope to be with you in about ten days. You will see my son sooner.

Joseph Priestley to Benjamin Vaughan: 30 July 1794
[Warrington Mss.]

Northumberland
July 30, 1794

Dear Sir,

I thought that you would rather chuse to hear from me at this time, than immediately on my arrival in this country, when I could not have given you any information from observations of my own. I have now seen all the principal people, and also persons who may be said to be in the *opposition*. I take no part in the politics of the country, and consort chiefly with your brother, and his friends, who are warm friends of *government*, as the phrase would be in England. I perceive, however, that the opposition is very considerable, and I am persuaded does not consist, as your brother will have it, of ill intentioned men. They are called *Antifederalists*, and object principally to the *excise laws*, and *funding system*, founded on a *national debt*, which they wish to have discharged, while others avow a liking of it, as a means of creating a dependence on the governing powers, which they think is wanting in this country, tho it has grown to a dangerous excess in England. The introduction of *paper money* is objected to, as having been the means of raising the price of every thing in the great towns so as to make living in them more expensive than in London. These enormous prices have not yet extended far into the country, but they must in time; and the rise having been sudden, and continually increasing, is certainly alarming. It will put a stop to all emigration, except of labourers, and make manufacturing hazardous. A riot has already been occasioned by the excise, and I fear more mischief will arise from it.—All parties agree in a wish for *peace with England*, but the opposition lay less stress upon it. Much is expected from the negotiation of Mr. Jay, but more from the success of the Arms of France, from which alone they expect a permanent peace. There is no doubt with any body, but that the Indians, who threaten a serious war, are instigated, and supplied with ammunition, by the English; and such is the impatience of the back settlers to hostility that they can hardly be restrained. The Vermontese want only leave to take all the forts in possession of the English themselves, and the people of Kentucky to open the Mississippi. They want no assistance from the general government. I hope that for once the English court will be wise enough to give reasonable satisfaction to the Americans, and that then a war will be prevented.

One of the worst things in this country, and what I did not expect to find (page torn) . . . is that the *poor laws* are the same as in England; and at New York and Philadelphia they already begin to find the same inconvenience from them. In Philadelphia the poor rate amounts to nine thousand pounds. Indeed, in many other things they seem to copy the English too closely, when they ought rather to take warning by the example.

For the sake of cheapness of living, and leisure for my pursuits, I think to reside chiefly here, visiting Philadelphia in the winter, as I used to do London. I shall, I think, prefer the climate of this country to that of England; but I am not yet reconciled to the different mode of living. But I never professed to leave England from *choice*, and from what has taken place there since I left it, I cannot but rejoice that I am where I am. I wish more of my friends were with me. Here I see the news only once a week, and the last we had was of the defeat of the Duke of York, and the suspension of the Habeas Corpus, the commitment of Mr. Stone (for whom I feel much interested) and the sending Mr. Tooke &c to the tower. These are considered here as desperate measures, and a prelude to some great convulsion, which I dread.

I shall not get my library, or Apparatus, built nor indeed absolutely fix the place of my residence till the next spring.

With every good wish to yourself, and all the family, in which we all join, I am, Dear Sir,

　　　Yours sincerely,
　　　J. Priestley
Addressed: "Benjn. Vaughan Esq. M. P.
　　No. 26 Finsbury Square,
　　London."
Endorsed: "Northumberland
　　July 30 1794
　　Dr. Priestley to B. V.
　　Recd. Oct. 6 1794."

William Vaughan to John Wilkinson: 25 October 1794
[Warrington Mss.]

Dear Sir

Mr. Lindsey is now with me he has a letter from Dr. P. of 14 Sepr. at Northumberland "What brought us here was the expectation of its being near the settlement that my Son & Mr. Cooper were projecting & behold *that is all over.* When the Lands came to be viewed they appeared not to be worth purchasing or accepting of–so that many will be sorely disappointed. They were deceived by the Proprietors & by the Evidence that had appearance of being satisfactory & the company had been put to much trouble–expence as well as delay. I now advise my Son to get a farm for himself near me & think no more of large partnerships which seldom answer. They have had several schemes but none seem likely to answer. What Mr. Cooper will do I cannot imagine. We must all live on our means for some Time to come.

The Professorship (sic) at the College of Philadelphia for Chemistry is supposed to be on his death bed. In case of a Vacancy Dr. Rush thinks the Dr. will be invited to succeed in this case I (sic: he) must reside 4 Months in the year at Philadelphia.

I have a letter from Mrs. Priestley long; cheerful; & gay; I sent it to the Ladies or would send at yours. She is contented & satisfied and likes the Country & people much. She does not feel so much the Evil of removing old Trees as she feared.

I think proper to give you this information as early as I could for your government both as interested and as a Man of Business. The disappointment may be great as unexpected to some who had prepared their funds or Who had built much on its success. If however it should be of any (use) to you–I could on the arrival of our fleet ease you a little of a heavy purse for 2 or 3 months for payment of duties &c. Subject to your own wishes in to command or (?) return you thus see. I have a ready expedient in converting Evils to benefits. Our Dash business has given me an aptness in converting all our Wants to one point.

> I am with great regard
> Dr Sir
> Yours sincerely
>   W. V.

Oct 25. 1794

Mr. Russell will be released & on Mr. Monroe's application.

Addressed: "John Wilkinson Esq
  Brosely
  Shifnal
  Shropshire
  to be forwarded if abroad or gone to Colebrooke Dale."
Endorsed: "Mr. W. Vaughan
  Octo 25 1794."

Joseph Priestley to John Adams: n.d. (March/April 1797)
[Mass. Hist. Soc. Colln., Adams Papers, Reel 383]

Dear Sir

Tho I shall always consider you as entitled to the greatest respect, as the chief magistrate of the country, I cannot help addressing you as a friend, who will feel for me in my situation, as I sincerely do for you in yours.

I have just received an account that my son in law is in fact become a bankrupt, and my daughter has five young children, and expects a sixth. My funds in this country do not furnish me with a sufficient resource for so great an emergency; but I have considerable property in France, the gift of a brother in law. This yields me nothing at present, but I have little doubt of making it productive if I were there. Now as I presume that you will soon send messengers or dispatches to France, could you favour me with a passage thither? In return, it might be in my power to render some service to this country with persons of influence in that; and this I should be happy in taking every opportunity of doing.

I am,
> Dear Sir,
> yours sincerely
> J Priestley

Tuesday morning

Benjamin Vaughan to (Charles) Vaughan: 25 September 1798
[A. P. S., Vaughan Papers, B. V. 46 p]

Hallowell, Septr. 25, 1798

My dear brother,

I can only say that Mr. S. had no invitation to write to me upon politics; & that I cannot be answerable for an act beyond my control. On the other hand, no letter or information whatever of any kind has passed from me to any one person on the continent of Europe, directly or indirectly, since I have been on the continent of America. I have been equally reserved as to political communications with persons in the British dominions; with the single exception of a very short passage written to one English friend, a number of months since. I have received but one letter from any person concerned in the French revolution, since my arrival in America: it referred to a religious subject, consisted of a single sentence, was free from politics, contained a reproach that none of my various friends on the continent of Europe had heard from me, & has itself remained without an answer. From the British dominions, I have not had one line upon politics, unless in favor of the British & American governments. Thus stands my political correspondence with Europe.

Since politics have become warm here, I have in the same proportion avoided them; & my political acts have been none; my writings have been none; and my conversations highly guarded. Among my limited acquaintance at this place, perhaps few have hitherto pretended to guess at my political sentiments; unless led to do so by others. If they have done so, it was contrary to my design. Besides, in this retired spot, with my retired life, & in agitated times, what is there possible to be done by an individual like myself?

When the proper authorities of this country shall inquire after me, which I presume they will not, I know what to reply. My recorded services to this country are, in the mean time, the protection on which I shall rely with others. I shall certainly avoid a newspaper correspondence on account of the sentiments of *other* persons, who it seems know so little of mine. I have no objection however to repeat, what I have so often intimated to you; that I have for some time renounced all politics, from the deep persuasion, that providence has special objects in view in the world at this time, which are of the most extensive & important nature; that its agents are chosen; its measures irresistible; & that the issue is to be waited for by persons like myself, with the most profound and tranquil submission.

You are right in supposing, that I have never rejoiced at the political sufferings of any. The awful acts of providence to which I refer, though deeply respected by me because its acts, yet fill me with an anguish inseparable from sentiments of humanity; & in this, I think I am justified by the sympathy repeatedly shewn by the author of our religion.—I will farther assure you, that in my present disposition of mind I dare rejoice at *nothing*, for I see the end of nothing.—We are in the hands of providence, whose ways are not our ways.

I write thus to set your mind at ease, & to prevent the few near you who know me from feeling pain on my account. For myself, I am content with a safe conscience. As I never asked or received any reward for my multiplied exertions for others, so I am perfectly resigned to whatever sufferings may be allotted to myself.

Yours ever affecty.

P.S. I have not seen the letter of Dr. P. to which you refer.

Joseph Priestley to Thomas Jefferson: 10 April 1801
[Lib. Cong. Mss., Jefferson Papers]

Philadelphia Apl 10. 1801

Dear Sir

Your kind letter, which, considering the numerous engagements incident to your situation, I had no right to expect, was highly gratifying to me, and I take the first opportunity of acknowledging it. For tho I believe I am completely recovered from my late illness, I am advised to write as little as possible. Your invitation to pay you a visit is flattering to me in the highest degree, and I shall not wholly despair of some time or other availing myself of it; but for the present I must take the nearest way home.

Your resentment of the treatment I have met with in this country is truly generous, but I must have been but little impressed with the principles of the religion you so justly commend, if they had not enabled me to bear much more than I have yet suffered. Do not suppose that, after the much worse treatment to which I was for many years exposed in England (of which the pamphlet I take the liberty to inclose will give you some idea) I was much affected by this. My *Letters to the Inhabitants of Northumberland* were not occasioned by any such thing, tho it served me as a pretence for writing them, but the threatenings of Mr. Pickering, whose purpose to send me out of the country Mr. Adams (as I conclude from a circuitous attempt that he made to prevent it) would not, in the circumstances in which he then was, have been able directly to oppose. My publication was of service to me in that and other respects, and I hope, in some measure, to the common cause. But had it not been for the extreme absurdity and violence of the late administration, I do not know how far these measures might not have been carried. Much, however, must be ascribed to the successes of the French and something also, perhaps, to the seasonable death of Genl. Washington. I rejoice more than I can express in the glorious reverse that has taken place, and which has secured your election. This I flatter myself will be the permanent establishment of truly republican principles in this country, and also contribute to the same desirable event in more distant ones.

I beg you would not trouble yourself with any answer to this. The knowledge of your good opinion and good wishes is quite sufficient for me. I feel for the difficulties of your situation, but your spirit and prudence will carry you thro them, tho not without paying the tax which the wise laws of nature have imposed upon predominance and celebrity of every kind, a tax which, for want of true greatness of mind, neither of your predecessors, if I estimate their characters aright, paid without much reluctance.

    With every good wish, I am,
       Dear Sir,
       yours sincerely
       J. Priestley

P.S. As I trust that *Politics* will not make you forget what is due to *science*, I shall send you a copy of some articles that are just printed for the *Transactions of the Philosophical society* in this place. No. 5 p 36 is the most deserving of your notice. I should have sent you my *defence of Phlogiston*, but that I presume you have seen it.

Joseph Priestley to Theophilus Lindsey, 19 December 1801
[A. P. S., Priestley Papers, B. P. 931]

No. 82. Northumberland Decr. 19. 1801
Dear friend

I rejoice with you on the unexpected return of *peace* after the most calamitous and destructive war. With this blessing I am happy to hear you have also that of *plenty*, having had, as we hear, a most abundant

harvest. May they long continue, and give relief to a suffering world. I
fear, however, that your difficulties will not end with the war, as there
must be a long account to make up, and consequently heavy taxes to be
imposed, in addition to those which the country is barely able to bear.
But I am confident that the greatest exertions will be made by those in
power to relieve the lower and middle classes, who will feel the burden
most, and almost all evils appear greater in prospect than they are found
to be in reality. I am a sincere well wisher to my native country, and shall
rejoice in every favourable prospect respecting it.

You will see by the message of our President to the Congress the
uncommonly pleasing prospect that is before us in this country. It is such
as was never made by the chief magistrate of any nation before. Having
made many savings of our usual expenditure, he proposes the abolishing
of *all internal taxes*; and yet promises the payment of our national debt
in much less time than the law, or any of the people, contemplated. He
reduces our very small standing army (which consisted I believe of not
more than 5000 men), but proposes to keep a fleet in the Mediterranean
to check the Barbary corsairs, which we are very well able to do without
any assistance from the European powers, to whose disgrace they have
continued their depradations so long. Instead of grasping at power, he
proposes that several things which had been left to the President should
be done by the Congress. It must, I should think, make his administration universally popular, great as the prejudices were that had been
raised against him by the opposite party.

I cannot say that the return of peace makes me look forward, as I once
did, to a visit to my friends in England. It is, on several accounts, too late
in my life for such an undertaking. My health received a rude shock from
the fever I had in Philadelphia, and several returns of the ague, from
which I am but lately recovered. My flesh and strength are greatly diminished, so that I should very ill bear the inconveniences of such a voyage,
and such a change in my mode of life. For having been long in a valetu-
dinary state, certain indulgences, which I can only find at home, are
become in a manner necessary to me. Besides the time in which I shall
be capable of doing any thing is now very short, and should not be
thrown away for the temporary satisfaction of a short interview with my
old friends, who are now much reduced in number. Seven years makes
a great change in the acquaintance of a person now near his 70th year.
You are by far the principal, and should I survive you, the world will
appear to me as a dreary waste, and I shall have little satisfaction except
in looking forward to a more happy meeting than I can contemplate here.
The slight mention you made of this circumstance brought tears into my
eyes. My wishes are the same with yours. But we are all at the disposal
of infinite wisdom and goodness, both in this world and another. My
study of the works of nature gives me a stronger feeling of this persuasion than I ever had before, and it increases with every days observations.
And yet there are philosophers who have no sentiment of the kind. I pity
them from my heart. It is the great joy of my life; but perhaps the more

in consequence of the many privations that I experience here tho I have abundant reason to be thankful for the circumstances in which I find myself.—I should do much better if I had a punctual bookseller; but the books Mr. Johnson should have sent early in the spring cannot now arrive before winter, all the ships expected from London being now come in, and none of them have brought me anything. He little thinks what I suffer for want of the books that he ought to send me, especially the Philosophical ones, as I am engaged in an important controversy, and do not hear of what is written for or against me till it is almost too late to profit by it. It would be the same with respect to theology, but that I am already possessed of almost every thing that I much want in that way. Thanks to your kind attention for it. I wish much to see the publication you commend so much of Mr. *Wm Belsham*, and also the *Lectures of his brother*; but I must wait at least half a year longer before I receive them. I am used to disappointments, but this is a severe one, and I know it is in vain to remonstrate.—Yours & Mrs. Lindsey most affectionately

     J Priestley

Addressed: "The Revd. Mr. Lindsey

    Essex Street

    London."

Endorsed: "four parcels previous to the box have been sent by various hands."

Joseph Priestley to Thomas Jefferson: n.d. (12 June 1802)
[Lib. Cong. Mss., Jefferson Papers]

To Thomas Jefferson, president of the united states of America
Sir,

My high respect for your character, as a politician, and a man, makes me desirous of connecting my name in some measure with yours while it is in my power, by means of some publication, to do it.

The first part of this work, which brought the history to the fall of the western empire, was dedicated to a zealous friend of civil and religious liberty, but in a private station. What he, or any other friend of liberty in Europe, could only do by their good wishes, by writing, or by patient suffering, you, Sir, are actually accomplishing, and upon a theatre of great and growing extent.

It is the boast of this country to have a constitution the most favourable to political liberty, and private happiness, of any in the world, and all say that it was yourself, more than any other individual, that planned and established it; and to this opinion your conduct in various public offices, and now in the highest, gives the clearest attestation.

Many have appeared the friends of the rights of man while they were subject to the power of others, and especially when they were sufferers by it; but I do not recollect one besides yourself who retained the same principles, and acted upon them, in a station of real power. You, Sir, have done more than this; having proposed to relinquish some part of the power which the constitution gave you; and instead of adding to the burdens of the people, it has been your endeavour to lighten those burdens tho the necessary consequence must be a diminution of your influence. May this great example, which I doubt not will demonstrate the practicability of truly republican principles, by the actual existence of a form of government calculated to answer all the useful purposes of government, (giving equal protection to all, and leaving every man in the possession of every power that he can exercise to his own advantage, without infringing on the equal liberty of others) be followed in other countries, and at length become universal.

Another reason why I wish to prefix your name to this work, and more appropriate to the subject of it, is that you have ever been a strenuous and uniform advocate of religious no less than civil liberty, both in your own state of Virginia, and in the united states in general; seeing in the clearest light the various and great mischiefs that have arisen from any particular form of religion being favoured by the state more than any other; so that the profession and practice of religion is here as free as that of philosophy, or medicine. And now the experience of more than twenty years leaves little room to doubt but that it is a state of things the most favourable to mutual candour, which is of great importance to domestic peace and good neighbourhood, and to the cause of all truth, religious truth least of all excepted. When every question is thus left to free discussion, there cannot be a doubt but that truth will finally prevail, and establish itself by its own evidence, and he must know little of mankind, or of human nature, who can imagine that truth of any kind will be ultimately unfavourable to general happiness. That man must entertain a secret suspicion of his own principles who wishes for any exclusive advantage in the defence or profession of them.

Having fled from a state of persecution in England, and having been exposed to some degree of danger in the late administration here, I naturally feel the greater satisfaction in the prospect of passing the remainder of an active life (when I naturally wish for repose) under your protection. Tho arrived at the usual term of human life it is now only that I can say I see nothing to fear from the hand of power, the government under which I live being for the first time truly favourable to me. And tho it will be evident to all who know me that I have never been swayed by the mean principle of fear, it is certainly a happiness to be out of the possibility of its influence, and to end ones days in peace, enjoying some degree of rest before the state of more perfect rest in the grave, and with the hope of rising to a state of greater activity, security, and happiness beyond it. This is all that any man can wish for, or have; and this, Sir, under your administration I enjoy.

With the most perfect attachment, and every good wish, I subscribe myself not your subject, or humble servant, but your sincere admirer.

J. Priestley

Joseph Priestley to Thomas Jefferson: 29 October 1802
[Lib. Cong. Mss., Jefferson Papers]

Dear Sir,

As there are some particulars in a letter I have lately received from Mr. Stone at Paris which I think it will give you pleasure to know, and Mr. Cooper has been so obliging as to transcribe them for me, I take the liberty to send them, along with a copy of my *dedication*, with the correction that you suggested, and a *Note* from the letter with which you favoured me concerning what you did with respect to the *constitution*, and which is really more than I had ascribed to you. For almost every thing of importance to political liberty in that instrument was, as it appears to me, suggested by you; and as this was unknown to myself, and I believe is so to the world in general, I was unwilling to omit this opportunity of noticing it.

I shall be glad if you will be so good as to engage any person sufficiently qualified to draw up such an account of the constitutional form of this country as my friend says will be agreeable to the emperor, and I will transmit it to Mr. Stone.

Not knowing any certain method of sending a letter to France and presuming that you do I took the liberty to inclose my letter to Mr. Stone. It is, however, so written, that no danger can arise to him from it, into whatever hands it may fall.

The state of my health, tho, I thank God, much improved, will not permit me to avail myself of your kind invitation to pay you a visit. Where ever I am, you may depend upon my warmest attachment and best wishes.

    Joseph Priestley

Northumberland Oct 29. 1802.

P.S. I send a copy of the *Preface* as well as of the Dedication, that you may form some idea of the work you are pleased to patronize.

Endorsed: "Priestley Joseph.Northumberld. Oct. 29. 1802. recd. Nov. 6."

Aurora: 15 February 1804

Extract of a Letter from Thomas Cooper, Esq. of Northumberland to James Woodhouse professor of Chemistry in the University of Pennsylvania, dated Feb. 6 1804.

Dear Sir,

Your old friend Dr. Priestley, died this morning without pain, at 11o'clock. He would have been 71 had he lived to the 24th of next month. He continued composed, cheerful and good tempered to the end. For some days he had been certain of his approaching fate.

P.S. Dr. P. desired his son to inform you that water imbibed 760 times its bulk of alkaline air.

BIBLIOGRAPHY

Manuscript Sources Cited

American Philosophical Society
Franklin Papers
Miscellaneous Manuscripts Collection
Priestley Papers
Vaughan Papers

Birmingham Public Libraries
Boulton and Watt Collection

Bowdoin College
Charles Vaughan Papers

Bowood Mss.
Vaughan Correspondence

Dickinson College
Priestley Collection

Dr. Williams's Library
Priestley-Lindsey Correspondence

Haverford College
Charles Roberts
Autograph Collection

Historical Society of Pennsylvania
Dreer Collection
Gratz Collection

Lea and Febiger Collection
Logan Papers
McKean Papers
Rush Mss.
Russell Papers

Library of Congress
Jefferson Papers

Liverpool Public Libraries
Roscoe Papers

Massachusetts Historical Society
Adams Papers

New York Public Library
Miscellaneous Papers, Charles Nisbet

Public Record Office
Treasury Solicitor's Papers

Royal Society
Priestley Mss.

Warrington Public Libraries
Priestley-Wilkinson Correspondence

Printed Sources Cited

Newspapers and Periodicals

Aurora
Cambridge Intelligencer
Gazette of the United States
Monthly Repository
Morning Chronicle
Northumberland Gazette

Philadelphia General Advertiser
Poulson's American Daily Advertizer
Porcupine's Gazette
Réimpression de l'Ancien Moniteur
  (Paris, 1858–63)

Published Works

Adams, C. F., ed., *Letters of John Adams addressed to his Wife* (Boston, 1841).
———, ed., *The Life and Works of John Adams* (Boston, 1850–56).
Adams, J. *A Defence of the Constitutions of Government of the United States of America* (London and Philadelphia, 1787).
———. *Discourses on Davila. A Series of Papers, on Political History. Written in the Year 1790, and then published in the Gazette of the United States. By an American Citizen* . . . (Boston, 1803).

187

Alger, J. G. "The British Colony in Paris, 1792-1793," *E. H. R.*, 13 (1898): 672-94.

American Philosophical Society: *Early Proceedings of the American Philosophical Society . . . compiled . . . from the Manuscript Minutes of its Meetings from 1744 to 1838* (Philadelphia, 1884).

Anderson, R. G. W. and Lawrence, C., eds., *Science, Medicine and Dissent: Joseph Priestley (1733-1804)* (London, 1987).

Anon, "A Brief Description of Joseph Priestley in a Letter of David English to Charles C. Green," *Presbyterian Historical Society Journal*, 38 (1960), 124-7.

Appleby, J. *Capitalism and a New Social Order. The Republican Vision of the 1790s* (New York, 1984).

———. "Republicanism in Old and New Contexts," *W. M. Q.*, 3rd Series, 43 (1986): 20-34.

———. "What is still American in the Political Philosophy of Thomas Jefferson?," *W. M. Q.*, 3rd Series, 39 (1982): 287-309.

Armytage, W. H. G. "The Editorial Experience of Joseph Gales, 1786-1794," *North Carolina Historical Review*, 28 (1951): 334-61.

Ashworth, J. "The Jeffersonians: Classical Republicans or Liberal Capitalists?" *Journal of American Studies*, 18 (1984): 425-35.

Bailyn, B. *Voyagers to the West. A Passage in the Peopling of America on the Eve of the Revolution* (New York, 1987).

Banning, L. "Jeffersonian Ideology Revisited: Liberal and Classical Ideas in the New American Republic," *W. M. Q.*, 3rd Series, 43 (1986): 3-19.

———. *The Jeffersonian Persuasion. Evolution of a Party Ideology* (Cornell Univ. Press, 1978).

Barlow, J. *Advice to the Privileged Orders in the Several States of Europe resulting in the Necessity and Propriety of a General Revolution in the Principle of Government* (London, 1792).

Barton, W. *Memoirs of the Life of David Rittenhouse* (Philadelphia, 1813).

Beard, C. A. *Economic Origins of Jeffersonian Democracy* (New York, 1913).

Belsham, T. *Memoirs of Theophilus Lindsey* (London, 1812).

Bernard, J. *Retrospections of America, 1797-1811* (New York, 1887, repr. 1969).

Beveridge, A. *The Life of John Marshall* (New York, 1916).

Bolton, H. C., ed., *Scientific Correspondence of Joseph Priestley* (New York, Privately Printed, 1892, repr. 1969).

Bonwick, C. *English Radicals and the American Revolution* (Chapel Hill, 1977).

———. "Joseph Priestley, Emigrant and Jeffersonian," *Enlightenment and Dissent*, 2 (1983): 3-22.

Boorstin, D. J. *The Lost World of Thomas Jefferson* (Boston, 1948).

Boyd, J. P., Bryan, R. M. and Hutter, E. L., eds., *The Papers of Thomas Jefferson*, Vol. 8 (Princeton Univ. Press, 1953).

Boyd, J. P. and Lester, R. W., eds., *The Papers of Thomas Jefferson*, Vol. 20 (Princeton Univ. Press, 1982).

Briggs, W. G. "Joseph Gales, Editor of Raleigh's First Newspaper," *North Carolina Booklet*, 7 (1907): 105-30.

Bromwell, W. J. *History of Immigration to the United States* (New York, 1856, repr. Arno, 1969).

Burke, E. *Reflections on the Revolution in France* (London, 1790).

Butterfield, L. H., ed., *Letters of Benjamin Rush* (Princeton Univ. Press, 1951).

Cappon, L. J., ed., *The Adams Jefferson Letters. The Complete Correspondence between Thomas Jefferson and Abigail and John Adams* (Chapel Hill, 1959).

Carter, E. C. "The Political Activities of Mathew Carey" (Bryn Mawr Ph.D. Thesis, 1962).

Chaloner, W. H. "Dr. Joseph Priestley, John Wilkinson and the French Revolution, 1789–1802," *Trans. Royal Hist. Soc.*, 5th Series, 8 (1958): 21–40.

Chinard, G. *Thomas Jefferson, The Apostle of Americanism* (2nd ed., revised, Univ. of Michigan Press, 1957).

———. *The Correspondence of Jefferson and Dupont de Nemours* (Johns Hopkins Univ. Press, 1931).

———. *Honest John Adams* (Boston, 1933).

———. "The American Philosophical Society and the World of Science (1768–1800)," *Proc. Am. Phil. Soc.* 87.1 (1944): 1–11.

———. "Jefferson and the American Philosophical Society," *Proc. Am. Phil. Soc.* 87.3 (1944): 263–76.

Clayden, P. W. *The Early Life of Samuel Rogers* (London, 1887).

Cobbett, W., ed., *The Parliamentary History of England . . . XXX* (London, 1817).

Cobbett, W. *Observations on the Emigration of Dr. Joseph Priestley, and on the Several Addresses delivered to him on his Arrival at New York* (Philadelphia, 1794).

———. *The Gros Mousqueton Diplomatique; or Diplomatic Blunderbus, containing Citizen Adet's Notes to the Secretary of State . . . With a Preface by Peter Porcupine* (Philadelphia, 1796).

———. *Remarks on the Explanation lately published by Dr. Priestley, respecting the intercepted letters of his friend and disciple, John H. Stone. To which is added a certificate of civism for Joseph Priestley, Jun.* (London, 1799).

———. *Porcupine's Works* (London, 1801).

Cohen, Seymour S. "Two Refugee Chemists in the United States, 1794: How We See Them," *Proc. Am. Phil. Soc.*, 126.4 (1982): 301–15.

Cohen, Sheldon S. "Thomas Wren: Ministering Angel of Forton Prison," *P. M. H. B.*, 103 (1979): 279–301.

Cole, G. D. H., ed., *The Life and Adventures of Peter Porcupine* (London, 1927).

Cole, G. D. H. *The Life of William Cobbett* (New York, 1924).

Coleridge, S. T. "Religious Musings," in E. H. Coleridge, ed., *The Complete Poetical Works of Samuel Taylor Coleridge* (Oxford, 1912).

Cooke, J. E., ed., *The Federalist* (Wesleyan Univ. Press, 1961).

Cooper, T. *Letters on the Slave Trade: first published in Wheeler's Manchester Chronicle; and since re-printed with additions and alterations* (Manchester, 1787).

———. *Propositions respecting the Foundation of Civil Government . . . read at the Literary and Philosophical Society of Manchester, on March 7, 1787, and first published in the Transactions of that Society, vol. 3. p. 481, 1790* (repr., Philadelphia, 1800).

———. *A Reply to Mr. Burke's Invective against Mr. Cooper and Mr. Watt in the House of Commons on the 30th of April 1792* (Manchester and London, 1792).

———. *Some Information Respecting America* (1st ed., London, 1794, 2d ed., London, 1795).

———. *Essays, originally inserted in the Northumberland Gazette, with Additions* (Northumberland, 1799).

———. *Political Essays, 2d. ed. with corrections and additions* (Philadelphia, 1800).

———. *Political Arithmetic* (Philadelphia, 1800).

———. *An Account of the Trial of Thomas Cooper, of Northumberland, on a charge of libel against the President of the United States; taken in short hand. With a Preface, Notes, and Appendix, by Thomas Cooper* (Philadelphia, 1800).

Corner, G. W., ed., *The Autobiography of Benjamin Rush: Mem. Am. Phil. Soc.*, 25 (Princeton Univ. Press, 1948).

Cunningham, N. E., Jr. *The Jeffersonian Republicans: The Formation of Party Organisation, 1789–1801* (Chapel Hill, 1957).

Currie, James (Jasper Wilson, pseud.), *A Letter, Commercial and Political, Addressed to the Rt. Hon. William Pitt, in which the Real Interests of Britain in the Present Crisis, are Considered, and some Observations are offered on the General State of Europe* (London, 1793).

Currie, W. W. *Memoir of the Life, Writings and Correspondence of James Currie* (London, 1831).

Dangerfield, G. *Chancellor Robert R. Livingston of New York (1746–1813)* (New York, 1960).

Davis, R. B., ed., *Jeffersonian America. Notes on the United States of America . . . by Sir Augustus John Foster, Bart.* (California, 1954).

Davis, R. W. *Dissent in Politics, 1780–1830. The Political Life of William Smith, M. P.* (London, 1971).

De Beaufort, R. L. *Memoirs of the Prince de Talleyrand* (London, 1891, repr. New York, 1973).

Durey, M. "Thomas Paine's Apostles: Radical Emigrés and the Triumph of Jeffersonian Republicanism," *W. M. Q.*, 3rd Series, 44 (1987): 661–88.

———. "Transatlantic Patriotism: Political Exiles and America in the Age of Revolutions," in C. Emsley and J. Walvin, eds., *Artisans, Peasants and Proletarians, 1760–1860* (London, 1985).

Duyckinck, E. A. and G. L. *Cyclopaedia of American Literature; embracing Personal and Critical Notices of Authors, and Selections from their Writings. From the Earliest Period to the Present Day* (New York, 1855–6).

Earl, J. L., "Talleyrand in Philadelphia, 1794–1796," *P. M. H. B.*, 91 (1967): 282–98.

Elkins, S. and McKitrick, E. *The Age of Federalism. The Early American Republic, 1788–1800.* (O. U. P., 1993).

Elliott, M. *Partners in Revolution. The United Irishmen and France* (Yale Univ. Press, 1982).

Foner, P. S. *The Democratic–Republican Societies 1790–1800* (Greenwood Press, 1976).

Ford, E. *David Rittenhouse, Astronomer-Patriot, 1732–1796* (Philadelphia, 1946).

Ford, P. L., ed., *The Works of Thomas Jefferson* (New York, 1904–5).

Fruchtman, J., Jr. "The Apocalyptic Politics of Richard Price and Joseph Priestley: A Study in Late Eighteenth-Century English Republican Millennialism," *Trans. Am. Phil. Soc.*, 73.4 (1983).

Garrett, C. *Respectable Folly: Millennarians and the French Revolution in France and England* (Johns Hopkins Univ. Press, 1975).

———. "Joseph Priestley, the Millennium, and the French Revolution," *Journal of the History of Ideas*, 34 (1973): 51–66.

Geffen, E. M. *Philadelphia Unitarianism, 1796–1861* (Univ. of Pennsylvania Press, 1961).

George, M. D. *Catalogue of Political and Personal Satires preserved in the Department of Prints and Drawings in the British Museum*, Vol. VII. (London, 1942).

Gerrald, J. *A Convention the only Means of Saving us from Ruin. In a Letter addressed to the People of England* (London, 1794).

Gibbs, F. W. *Joseph Priestley. Adventurer in Science and Champion of Truth* (London, 1965).

Godwin, W. *Enquiry Concerning Political Justice* (London, 1793).

Goodwin, A. *The Friends of Liberty. The English Democratic Movement in the Age of the French Revolution* (Harvard Univ. Press, 1979).

Graham, J. "Revolutionary Philosopher: The Political Ideas of Joseph Priestley (1733–1804), Part One," *Enlightenment and Dissent*, 8 (1989): 43–68; and "Part Two," *Enlightenment and Dissent*, 9 (1990): 14–46.

——. "A Hitherto Unpublished Letter of Joseph Priestley," *Enlightenment and Dissent*, forthcoming.

——. *The Nation, the Law and the King: Reform Politics in England, 1789–99*, Univ. Press of America, forthcoming.

Green D. *William Cobbett, The Noblest Agitator* (London, 1983).

Hansen, M. L. *The Atlantic Migration, 1607–1860* (Harvard Univ. Press, 1940).

Haraszti, Z. *John Adams and the Prophets of Progress* (Harvard Univ. Press, 1952).

Hindle, B. *David Rittenhouse* (Princeton, 1964).

Horsman, R. "The British Indian Department and the Resistance to General Anthony Wayne, 1793–1795," *Mississippi Valley Historical Review*, 49 (1962): 269–90.

Howe, J. R., Jr. *The Changing Political Thought of John Adams* (Princeton Univ. Press, 1966).

Howell, T. B. and T. J., eds., *A Complete Collection of State Trials* (London, 1809–1828).

Jefferson, T. *Notes on the State of Virginia*, in M. D. Peterson, ed., *Thomas Jefferson* (New York, 1984).

Jeremy, D. J., ed., "Henry Wansey and his American Journal, 1794," *Mem. Am. Phil. Soc.*, 82 (Philadelphia, 1970).

Jeyes, S. H. *The Russells of Birmingham in the French Revolution and America, 1791–1814* (London, 1911).

Kennedy, M. L. *The Jacobin Clubs in the French Revolution. The First Years* (Princeton, 1982).

Knight, F. *The Strange Case of Thomas Walker* (London, 1957).

Koch, A. *The Philosophy of Thomas Jefferson* (New York, 1943).

Koch, A. and Ammon, H. "The Virginia and Kentucky Resolutions: An Episode in Jefferson's and Madison's Defense of Civil Liberties," *W. M. Q.*, 3rd Series, 5 (1948): 145–76.

Kramnick, I. "Eighteenth-Century Science and Radical Social Theory: The Case of Joseph Priestley's Scientific Liberalism," *Journal of British Studies*, 25 (1986): 1–30.

——. "Republican Revisionism Revisited," *Amer. Hist. Rev.*, 87.3 (1982): 629–64.

——. *Republicanism and Bourgeois Radicalism. Political Ideology in late Eighteenth Century England and America* (Cornell Univ. Press, 1990).

La Rochefoucauld Liancourt, F.A.F., Duc de, *Travels through the United States of America* (London, 1799).

Link, E. P. *The Democratic-Republican Societies, 1790–1800* (Columbia Univ. Press, 1942).

McCoy, D. *The Elusive Republic. Political Economy in Jeffersonian America* (Univ. of N. Carolina Press, 1980).

McLachlan, H. *The Letters of Theophilus Lindsey* (Manchester, 1920).

Malone, D. *The Public Life of Thomas Cooper (1783–1839)* (New Haven, Conn., 1926).

——. *Jefferson and the Ordeal of Liberty* (Boston, 1962).

Malthus, T. R. *An Essay on the Principle of Population* (London, 1807 ed.).

Marvin, M. V. *Benjamin Vaughan* (Privately Printed, 1979).

Miller, J. C. *The Federalist Era: 1789–1801* (New York, 1960).

Miller, P. N., ed., *Joseph Priestley. Political Writings* (C. U. P., 1993).

Miller, W. "The Democratic Societies and the Whiskey Insurrection," *P. M. H. B.*, 62 (1938), 324–49.

Muir, T. *An Account of the Trial of Thomas Muir . . . for Sedition* (Robertson's edition, Edinburgh, 1793).

Murray, C. *Benjamin Vaughan (1751–1835). The Life of an Anglo-American Intellectual* (New York, 1982).

Oberg, B. et al., eds., *The Papers of Benjamin Franklin*, Vols 29, 30 (Yale Univ. Press, 1992, 1993).

Paine, T. *Rights of Man, Parts One and Two* (London, 1791, 1792).

———. *The Age of Reason; being an investigation of true and fabulous theology* (London, 1795).

———. *Agrarian Justice, opposed to Agrarian Law, and Agrarian Monopoly. Being a plan for ameliorating the condition of man, by creating in every nation a National Fund* (London, 1797).

Paley, W. *A View of the Evidences of Christianity* (London, 1794).

Park, M. C. "Joseph Priestley and the Problem of Pantisocracy," *Proceedings of the Delaware County Institute of Science*, 11.1 (1947): 1–60.

Peeling, J. H. "Governor McKean and the Pennsylvania Jacobins, 1799–1808," *P. M. H. B.*, 54 (1930): 320–54.

Peterson, M. D. *Thomas Jefferson and the New Nation. A Biography* (O. U. P., 1970).

———. *Adams and Jefferson. A Revolutionary Dialogue* (Univ. of Georgia Press, 1976).

Pocock, J. G. A. "Religious Freedom and the Desacrilization of Politics: From the English Civil Wars to the Virginia Statute," in M. D. Peterson and R. C. Vaughan, eds., *The Virginia Statute for Religious Freedom. Its Evolution and Consequences in American History* (C. U. P., 1988).

Powell, W. S., ed., "The Diary of Joseph Gales, 1794–1795," *North Carolina Historical Review*, 26 (1949): 335–47.

Prelinger, C. M. "Benjamin Franklin and the American Prisoners of War in England during the American Revolution," *W. M. Q.*, 3rd Series, 32 (1957): 261–94.

Price, R. *A Discourse on the Love of our Country* (London, 1789).

Priestley, J. *An Essay on the First Principles of Government and on the Nature of Political, Civil, and Religious Liberty . . .* (London, 1768).

———. *The Present State of Liberty in Great Britain and her Colonies* (London, 1769).

———. *An Address to Protestant Dissenters of all Denominations, on the Approaching Election of Members of Parliament* (London, 1774).

———. *An History of the Corruptions of Christianity* (Birmingham, 1782).

———. *The Importance and Extent of Free Enquiry in Matters of Religion. A Sermon preached before the Congregations of the Old and New Meetings of Protestant Dissenters at Birmingham, Nov. 5, 1785. To which are added, Reflections on the Present State of Free Enquiry in this Country* (Birmingham, 1785).

———. *A Letter to the Rt. Hon. William Pitt . . . on the Subjects of Toleration and Church Establishments* (London, 1787).

———. *Lectures on History and General Policy* (Birmingham, 1788).

———. *Familiar Letters, Addressed to the Inhabitants of Birmingham* (Birmingham, 1790).

———. *A General History of the Christian Church to the Fall of the Western Empire* (Birmingham, 1790).

———. *Letters to the Rt. Honourable Edmund Burke, Occasioned by his Reflections on the Revolution in France* (Birmingham, 1791).

———. *The Proper Objects of Education in the Present State of the World, represented in a Discourse, delivered . . . to the supporters of the New College at Hackney* (London, 1791).

———. *A Political Dialogue on the General Principles of Government* (London, 1791).

———. *Letter to the Inhabitants of Birmingham, Morning Chronicle*, 20 July 1791.

———. *An Appeal to the Public on the subject of the Riots in Birmingham*, Parts I and II (London, 1791, 1792).

———. *The Present State of Europe compared with Ancient Prophecies; A Sermon preached at the Gravel-Pit Meeting, in Hackney, February 28, 1794 . . . with a Preface, containing the Reasons for the Author's leaving England* (London, 1794).

———. *The Use of Christianity, especially in difficult Times. A Sermon delivered at the Gravel-Pit Meeting, Hackney, March 30 1794, being the Author's Farewell Discourse to his Congregation* (London, 1794).

———. *Letters to a Philosophical Unbeliever . . . containing an Answer to Mr. Paine's Age of Reason* (Philadelphia, 1795).

———. *Observations on the Increase of Infidelity. First prefixed to the American edition of the Letters to the Philosophers and Politicians of France, but now much enlarged* (Northumberland, Pa., 1795, repr. London, 1796).

———. *Observations on the Increase of Infidelity . . . The 3rd ed. to which are added, Animadversions on the Writings of several modern unbelievers, and especially the Ruins of Mr. Volney* (Philadelphia, 1797).

———. *Discourses relating to the Evidences of Revealed Religion* (Philadelphia, 1796, 1797).

———. *The Case of the Poor Emigrants Recommended in a Discourse delivered at the University Hall in Philadelphia . . . February 19, 1797* (Philadelphia, 1797).

———. *Maxims of Political Arithmetic, applied to the Case of the United States of America* (Philadelphia, 1798).

———. *Letters to the Inhabitants of Northumberland and its Neighbourhood, on subjects interesting to the author, and to them* (1st ed., Northumberland, Pa., 1799).

———. *A Comparison of the Institutions of Moses with those of the Hindoos and other Ancient Nations* (Northumberland, 1799).

———. *The Doctrine of Phlogiston established and that of the Composition of Water refuted* (Northumberland, 1800).

———. *Letters to the Inhabitants of Northumberland . . . The 2d ed. with additions, to which is added a letter to a friend in Paris, relating to Mr. Liancourt's Travels in the North American States* (Philadelphia, 1801).

———. *A General History of the Christian Church from the Fall of the Western Empire to the present time* (Northumberland, 1802).

———. "Experiments on the Production of Air by the Freezing of Water," *Transactions of the American Philosophical Society*, 5 (1802): 36–41.

———. *Lectures on History and General Policy; to which is prefixed, an Essay on a Course of Liberal Education for Civil and Active Life: and an Additional Lecture on the Constitution of the United States* (Philadelphia, 1803).

———. *Memoirs of Dr. Joseph Priestley to the year 1795 written by himself with a continuation to the time of his decease, by his son, Joseph Priestley: and Observations on his Writings, by Thomas Cooper President Judge of the 4th District of Pennsylvania, and the Rev. William Christie* (Northumberland, 1806, repr. New York, 1978).

———. *Notes on all the Books of Scripture, for the Use of the Pulpit and Private Families* (Northumberland, 1803–4).

Robbins, C. "Honest Heretic: Joseph Priestley in America, 1794–1804," *Proc. Am. Phil. Soc.* 106.I (1962): 60–76.

Roberts, K. and A. M., eds., *Moreau de St. Méry's American Journey (1793–1798)* (New York, 1947).

Rose, R. B. "The Priestley Riots of 1791," *Past and Present*, 8 (1960): 68–88.

Rutland, R. A., Mason, T. A., et al., eds., *The Papers of James Madison* (Univ. Press of Virginia, 1983).

Rutt, J. T., ed., *The Theological and Miscellaneous Works of Joseph Priestley* (London, 1817–31) I. Parts 1 and 2: *Life and Correspondence.*

Sawvel, F. B., ed., *The Complete Anas of Thomas Jefferson* (New York, 1903).

Schofield, R. E., ed., *A Scientific Autobiography of Joseph Priestley (1733-1804)* (Cambridge, Mass., and London, 1966).

Schofield, R. E. *The Lunar Society of Birmingham; a Social History of Provincial Science and Industry in Eighteenth Century England* (Oxford, Clarendon, 1963).

Schwartz, T. A. and McEvoy, J. G., eds., *Motion Toward Perfection: The Achievement of Joseph Priestley* (Boston, 1990).

Scott, F. D. *The Peopling of America: Perspectives on Immigration* (Washington, American Historical Association, 1984).

Sheps, A. "Ideological Immigrants in Revolutionary America," in P. Fritz and D. Williams, eds., *City and Society in the Eighteenth Century* (Toronto, 1973).

Slaughter, T. P. *The Whiskey Rebellion. Frontier Epilogue to the American Revolution* (O. U. P., 1986).

Smith, A. *An Inquiry into the Nature and Causes of the Wealth of Nations*, R. H. Campbell and A. S. Skinner, eds. (Oxford, Clarendon, 1976).

Smith, J. A. *Franklin and Bache. Envisioning the Enlightened Republic* (O. U. P., 1990).

Smith, J. M. *Freedom's Fetters: the Alien and Sedition Laws and American Civil Liberties* (Cornell Univ. Press, 1956).

———. "The Grass Roots Origins of the Kentucky Resolutions," *W. M. Q.*, 3rd Series, 27 (1970): 221-45.

Smith, P. *John Adams* (New York, 1962).

Stetson, S. P. "The Philadelphia Sojourn of Samuel Vaughan," *P. M. H. B.*, 73 (1949): 459-74.

Stone, J. H. *Copies of Original Letters recently written by Persons in Paris to Dr. Priestley in America. Taken on board of a neutral Vessel* (2nd ed., London, 1798).

Stuart, D. *Peace and Reform, against War and Corruption . . .* (London, 1794).

Syrett, H. C. and Cooke, J. E., eds., *The Papers of Alexander Hamilton* (Columbia Univ. Press, 1966).

Tagg, J. *Benjamin Franklin Bache and the Philadelphia Aurora* (University of Pennsylvania Press, 1991).

Thorpe, T. E. *Joseph Priestley* (London, 1906).

Tolles, F. B. *George Logan of Philadelphia* (New York, O. U. P., 1953).

Turner, E. R. "The Abolition of Slavery in Pennsylvania," *P. M. H. B.*, 36 (1912): 129-42.

Turner, F. J., ed., "Correspondence of the French Ministers to the United States, 1791-1797," *Annual Report of the American Historical Association . . . 1903*, II (Washington, 1904, repr. New York, 1972).

Twomey, R. J. "Jacobins and Jeffersonians: Anglo-American Radical Ideology, 1790-1810," in M. and J. Jacob, eds., *The Origins of Anglo-American Radicalism* (London, 1984).

Tyson, G. P. *Joseph Johnson. A Liberal Publisher* (University of Iowa Press, 1979).

Vaughan, B., ed., *Political, Miscellaneous, and Philosophical Pieces . . . written by Benjamin Franklin* (London, 1779).

(Vaughan, B.) *Letters on the Subject of the Concert of Princes and the Dismemberment of France and Poland* (London, 1793).

(Vaughan, B.) *Comments on the proposed War with France, on the State of Parties, and on the New Act respecting Aliens . . .* (London, 1793).

(Vaughan, B.) *De l'Etat Politique et Economique de la France sous la Constitution de l'An III* (Strasbourg, 1796).

Wakefield, G. *A Reply to some Parts of the Bishop of Llandaff's Address* (London, 1798).

Webster, N., Jr., *Ten Letters to Dr. Priestley, in Answer to his Letters to the Inhabitants of Northumberland* (New Haven, 1800).

Wilbur, E. M. *A History of Unitarianism in Transylvania, England, and America* (Harvard Univ. Press, 1952).

Willcox, W. B., ed., *The Papers of Benjamin Franklin*, Vol. 16 (Yale Univ. Press, 1972).

Williams, H. M. *A Tour in Switzerland, or, a View of the present State of the Government and Manners of these Cantons, with comparative Sketches of the Present State of Paris* (London, 1798).

Woodward, L. D. *Hélène Maria Williams et ses Amis* (Paris, 1930, repr. Genève, 1977).

INDEX

Abercrombie, Rev. Dr., 33 n.

Adams, Abigail, 90 n.; 95 n.; 123; 125 and n.

Adams, John, 61, 80 n., 87 n., 90 n., 92, 99, 112, 114, 123, 148 n., 154; on European emigrants, 3; Ambassador to England, meets Priestley, 13–16; Priestley to on persecution, emigration, 28, 172; Adams condoles with, sends Priestley his *Defence*, assures Priestley of welcome in America, 28 n., 40, 171–2; suggests settlement in Boston, 48 and n.; Priestley to introducing his sons, 33 n.; disclaims interest in politics, 39, 64, 79 and n., 173; Adams's admiration for English Constitution, distrust of French philosophes, 40; as President passes legislation repressing freedom of opinion, 40, 107–8; Priestley to on French revolution, on preference for America to England, 45, 172–3; Adams defines republicanism, 45–6: *Discourses on Davila*, 45–6, *Defence of Constitutions*, 45; opposition to French revolution, 45; monarchy might have to be re-established in America, 46, 168; Priestley sends pamphlets to, 60; writes to on failure of settlement, 64; on conflict in Europe, 65, 68; Adams offers to help Priestley with post, 76; writes to on naturalisation laws (1795), 78–9 and n.; attends Priestley's preaching in Philadelphia, 89; Priestley dedicates *Discourses* to, 89 and n., 100; Adams's policy towards France, 93–4; differences with Jefferson, 94, 106; election as President, 95; skeptical of Priestley's views on French Revolution, millennium and prophecies, 95 and n.; fails to attend second set of *Discourses*, 95; fails to reply to appeal for passage to France, 97, 125–6 n., 179; ignores application for government office for Cooper, 99, 118, 123; fails to subscribe to *Church History*, 100; introduces legislation preventing emigration, suppressing political opposition, 107–8; replies to Patriotic Addresses, denounces French expansionism and militarism, 107; Priestley criticizes warlike measures of, 109; liable to be deported by, 41, 109–10, 133, 137–2, 146–7 and n.; criticizes abuse of executive power by, 109–10, 117; in *Letters*, 134, 137–40; incensed by illiberal measures of, 118; Cooper's activities against, 118; low opinion of, 118–19; attacks in *Northumberland Gazette*, 119–22;

in *Aurora* and Pennsylvania governorship election, 123; Adams convinced of necessity of prosecuting Cooper, 123; Cooper convicted of a libel upon, 123, 162 n.; Adams informed of Priestley's political activities, 123–4; resists calls for deportation, on lack of influence, advises to avoid politics, 125–6 and n., 133, 146–7 and n., 181; Jefferson opposes measures of, 94, 106–8, 116 n., 140–4, 147–8, 167; Priestley believes republican principles in great danger under, 146, 149, 181; Adams on Priestley's character, 165–6; retrospectively, on Priestley's influence, 166

Adams, John Quincy, "Publicola letters," 46; attacks Priestley and Cooper in *Gazette of United States*, 146

Adams, Thomas Boylston, 125

Adet, Pierre Auguste, French Minister in Philadelphia (1795–7), appeals in *Aurora*, 94, 98 n.; acquaintance with Priestley, 97–8, 105, 106 n., 113; scientific controversy with, 98 n.; departure for France, 98

Admiralty, London, "Intercepted letters" exhibited in, 110 n.

Aikin, Lucy, 7 n.

Air, Priestley's experiments on, at Bowood, 10; in Northumberland, 78, 181, 185

Albany, edition of Priestley's *Letters* published in, 140–1

Alexander I of Russia, J. H. Stone's account of, asks for description of American Constitution for, Priestley's interest and Jefferson's reply, 156 and n.

Agriculture

In America: Cooper practices in Northumberland, 81–2, 84, 86–7; urges priority be given to, 103 n., 119–20 and n; Jefferson on overriding importance of, 92 and n., 93 and n.; welcomes Cooper's and Priestley's writings on, 120 n., 142 and n.; as President declares commerce necessary for disposing of products of, 153; Logan as propagandist for, 92–3 n.; Priestley's sons' practice, 29 n., 80–1, 178; Priestley urges priority over commerce in America, 103 and n., 120 n.; advocates equal treatment with commerce, 161

In England: Adam Smith's views on, 12, 104 n., 120 n.; Priestley on role of in society 12–13

In France: Hurford Stone on, 112; influence of du Pont de Nemours on Jefferson, 143–4

www.ingramcontent.com/pod-product-compliance
Lightning Source LLC
Chambersburg PA
CBHW080924100426
42812CB00007B/2360